Deploying and Managing
IP over WDM Networks

For a complete listing of the *Artech House Telecommunications Library,*
turn to the back of this book.

Deploying and Managing IP over WDM Networks

Joan Serrat
Alex Galis

Editors

Artech House
Boston • London
www.artechhouse.com

Library of Congress Cataloging-in-Publication Data

A catalog record of this book is available from the Library of Congress

British Library Cataloguing in Publication Data

A catalog record of this book is available from the British Library

ISBN 1-58053-501-1

Cover design by Gary Ragaglia

© 2003 ARTECH HOUSE, INC.
685 Canton Street
Norwood, MA 02062

International Standard Book Number: 1-58053-501-1

10 9 8 7 6 5 4 3 2 1

Contents

Foreword

Over the past 25 years, the global telecommunications infrastructure has been transformed. The installation of millions of miles of optical fiber cables has led to orders-of-magnitude increases in bandwidth and capacity. The availability of this greatly increased capacity in the global core/backbone networks, and the resultant large cost reductions for bandwidth have opened the way for the huge growth in Internet services over the last 10 years. The development of mass market mobile telecommunications has contributed an extra dimension. These developments, combined with the growth of broadband access, are driving the demand for even higher speed, better network performance, enhanced functionality, and lower cost.

Further advances in network performance and functionality will come from the integration of, and synergies between, traditionally separate voice, video, and data networks. This will be true not only in the equipment with which the network is constructed, but particularly in the network operation, administration, and management. As networks grow in size and functionality, the challenge of network management and control becomes both more complex and more important. The coherent management of network resources in a multidomain multienvironment is fundamental in supporting the widespread deployment of new telecommunications networks and services in a competitive and ever-changing market.

Research on these key issues has been well supported within the European Union's Information Society Technologies (IST) program (part of the Fifth Framework Program for Research and Technological Development). The goals of the IST program address research and development of advanced communications systems and services for economic development and social cohesion in

Europe, taking into account the rapid evolution in technologies, the changing regulatory situation, and the opportunities for development of advanced trans-European networks and services. The research supports European policies for the early deployment and effective use of advanced communications, including broadband, in consolidation of the internal market and the enabling of European industry to compete effectively in the global marketplace.

In the areas of computing, communications, and networks, the IST program has fostered and supported research and development in projects covering a wide range of essential technologies and systems, including network integration, interoperability and interworking, technologies for network management and service-level interworking, and terabit multiwavelength optical networks. The work has been organized and implemented through clusters of projects, which are grouped together to exchange knowledge and results relating to common themes. This provides synergies and opportunities for better exploitation of the results of the work-programs of each individual project.

This book is based on the experiences of the researchers in the WDM and IP Network Management (WINMAN) project, with contributions from the LION and OPTIMIST projects. The book addresses the integration of the IP and the WDM domains and covers also key aspects of the deployment of these network technologies. It makes a significant contribution to advancing the state of the art in integrated communication management for IP over WDM, developing enabling mechanisms for end-to-end management solutions across multidomain and multitechnology networks. These management mechanisms enable interoperability between management domains, between heterogeneous network technology, and between service and network management functions in a multiprovider environment. It presents the background to network technologies and management systems, and as such it introduces the reader gradually to its main subjects. It also provides an outlook for the IP-over-WDM network architecture and management.

The book is intended as a reference for use by:

- Network and system managers, to determine the techniques and mechanisms for end-to-end management problems in the context of IP-over-WDM networks;
- Network management vendors, to understand the benefits and limitations of current practices and standards as applied to multidomain management problems;
- Students and researchers of advanced communications management techniques, to learn about important concepts and issues related to management of networks.

All of these people will find this book very useful, and I warmly recommend it as an important contribution to the subject. The views expressed in this article are those of the author and do not necessarily reflect the official European Commission's view on the subject.

Andrew Houghton
Area Coordinator for Optical Network Research
IST Program
Directorate-General Information Society
European Commission

Preface

The integration of the Internet with optical networks and management of heterogeneous and hybrid networks has been always a challenge for network and service operators. Different frameworks and architectural approaches have been proposed and investigated in the research literature and in the commercial world. The purpose of this book is to introduce the reader to the current state of the art and the future challenges of the integration of the IP and WDM networks, to present a novel management approach to hybrid IP over WDM networks and to present part of the results developed in three European Union research projects: IST-WINMAN, IST-LION, and IST-OPTIMIST.

This book has benefited greatly from the contributions, review, and experience of many people, who generously gave us their time and experience.

Moreover, the editorial tasks that were necessary to select and shape the material were a joint effort between the following persons and us:

Ann Ackaert, Ghent University;
Harold Balemans, Lucent Technologies;
Dimitris Chronis, OTE;
Javier González-Ordás, Telefónica I+D;
Eduardo Grampín-Castro, UPC;
Gerard Hoekstra, Lucent Technologies;
Fotis Karayannis, Temagon;
Harris Katopodis, OTE;

Eugene Kozlovsky, UCL;
Antonio Manzalini, TILAB;
Lampros Raptis, NTUA;
Anat Schwartz, TTI Telecom;
Nuno Silva, Portugal Telecom;
Kostas Vaxevanakis, Ellemedia
   Technologies.

This book is based on the results of work carried out by several research consortia and particularly by the following people listed as coauthors.

Coauthors of the IST-WINMAN project contribution:

Pablo Arozarena-Llopis, Telefónica I+D;
Theodore Avgeris, OTE-Consulting;
Harold Balemans, Lucent Technologies;
Javier Baliosian-De-Lazzari,
    Univ. Politècnica Catalunya;
George Bathas, OTE;
Polina Bayvel,
    University College London;
Giorgos Chatzilias,
    Nat.Tech. University Athens;
Dimitris Chronis, OTE
Luke Davis, TTI Telecom;
Alicia Díaz-Marlasca, Tecsidel
Joanna Edmonds, TTI Telecom;
Michalis Ellinas,
    Nat.Tech.University Athens;
Alex Galis, University College London;
Roberto García-Hernández,
    Univ.Politècnica Catalunya;
Javier González-Ordás, Telefónica I+D;
Eduardo Grampín-Castro,
    Univ.Politècnica Catalunya;
Abdelkader Hajjaoui,
    Lucent Technologies;
Gerard Hoekstra, Lucent Technologies;
Fotis Karayannis, Temagon;
Harris Katopodis, OTE
Eugene Kozlovsky,
    University College London;

Alexandre Laranjeira,
    Portugal Telecom Inovaçao;
José-Antonio Lozano, Telefónica I+D;
Fernando Martí-Pallarés, Tecsidel;
Bruno Mayersohn, TTI Telecom;
Rui Oliveira, Portugal Telecom Inovaçao;
George Pagomenos,
    Ellemedia Technologies;
Yiorgos Patikis,
    Nat.Tech.University Athens;
Panos Philippopoulos, Temagon;
Lampros Raptis,
    Nat.Tech.University Athens;
Willem Romijn, Lucent Technologies;
Javier Rubio-Loyola,
    Univ.Politècnica Catalunya;
Anat Schwartz, TTI Telecom;
Joan Serrat-Fernández,
    Univ.Politècnica Catalunya;
Nuno Silva, Portugal Telecom Inovaçao;
Filipe Sousa, Portugal Telecom Inovaçao;
Gijs Van Ooijen, Lucent Technologies;
Nikos Vardalachos,
    University College London;
Konstantinos Vaxevanakis,
    Ellemedia Technologies;
Rodrigo Werlinger-Cruces,
    Telefónica CTC;
Theodore Zahariadis,
    Ellemedia Technologies.

Coauthors of the IST-LION project contribution:

Didier Colle, IMEC;
Sophie De Maesschalck, IMEC;
Piet Demeester, IMEC;
Ralf Geerdsen, T-Systems;
Ulrike Hartmer, T-Systems;
Georg Lehr, T-Systems;
Ilse Lievens, IMEC;
Johann Maierhofer, SICN;

Antonio Manzalini, TILAB;
Roberto Morro, TILAB;
Uwe Pauluhn, SICN;
Marco Quagliotti, TILAB;
Giuseppe Ricucci, TILAB;
Josep Sole-Pareta, UPC;
Salvatore Spadaro, UPC;
Rafal Stankiewicz, AGH.

Coauthors of the IST-OPTIMIST project contribution:

Ann Ackaert, Ghent-University/IMEC;
Didier Colle, Ghent-University/IMEC;
Piet Demeester, Ghent-University/IMEC;
Didier Erasme,
   Ecole National Sup. des Telecomm.;
Evguenii Guiorguiev,
   Ecole National Sup. des Telecomm.;
Paul Lagasse, Ghent-University/IMEC;
Christophe Minot,
   Ecole National Sup. des Telecomm.;

Mike O'Mahony, University of Essex;
Erwin Patzak, Heinrich Hertz Institut;
Tanya Politi, University of Essex;
Juergen Saniter, Heinrich Hertz Institut;
Bjarne Tromborg, Research Center COM;
Paul Vogel, Telscom Consulting;
Tommy Winter Berg,
   Research Center COM.

This book could not have been completed without the enthusiastic support of every individual listed here. We thank all contributors and invite you to consider and make use of the concepts and technological results presented in this book.

Acknowledgments

The editors would like to acknowledge the project IST-WINMAN, partially funded by the Commission of the European Union, that allowed us to go forward with the research in this area.

We thank all contributors to this book, especially to Harold Balemans, the IST-WINMAN project manager, for his contributions and management of the project.

Special thanks to Ann Ackaert (Ghent-University) and Antonio Manzalini (TILAB) for their leading activity making possible the contributions from the IST-OPTIMIST and IST-LION projects, respectively.

We would like to acknowledge the help and enthusiastic support received from Julie Lancashire and Tiina Ruonamaa, the two Artech House officers who conducted the book review and publication process.

We thank Professor Chris Todd (UCL) for his support and encouragement for writing this book.

We thank Melchor Fuentes (UPC) for his indispensable logistical support and artistic skills, and Taina Galis for her helpful comments on improving the readability of the book.

Finally, we would like to thank Andrew Houghton, European Union Project coordinator, Hien Nguyen Quang (Siemens), Nico Wauters (Tellium), and Sylvain Grisouard (Alcatel), WINMAN project reviewers, for their support, wisdom, and encouragement for the work of the WINMAN project. They have modulated the evolution of the project and therefore affected the content of this book.

1

Introduction

1.1 The Importance of IP and WDM Networks

1.1.1 Historical Perspective

Historically, communications networks could be divided into different segments: local area networks (LANs) and wide area networks (WANs), including the backbone networks. The IP dominated the LANs area, where all of the applications feed the users. The synchronous optical network/synchronous digital hierarchy (SONET/SDH) has always been strong in the WANs area, specifically in the core of the network to multiplex traffic (mainly voice traffic), while asynchronous transfer mode (ATM) was used in the WANs for switched traffic. In most cases, ATM worked on top of SDH where traffic demands were higher than the minimum SDH traffic rate (STM-1 or 155 Mbps) or on top of pleosynchronous digital hierarchy (PDH) for lower traffic rates.

The integration of multiple layers was soon a reality, forming stacks such as IP over ATM over SDH (IP/ATM/SDH) or IP over ATM over PDH (IP/ATM/PDH). In other words, most WAN traffic, including IP, is converted into ATM cells and then transported over SDH paths. This usually occurs because IP traffic needs to be multiplexed with ATM traffic or other time division multiplex (TDM) traffic to make the data transport cost effective.

1.1.2 Current Trends

The current telecommunications market evolved to be highly driven by IP-oriented applications and technologies, mainly caused by the explosive growth in the number of Internet users. The original focus on academic-oriented use has shifted to the development of new services used by the general public and commercial organizations. According to the survey of the Internet Software Consortium, in July 2002 there were well over 126 million hosts on the Internet.

This is a completely different situation with respect to what happened in the past, when almost all of the traffic of telecommunications networks was voice information from telephone calls. The technology that was foreseen only for specific data communication is being pushed to be of widespread use. In addition to data-oriented applications, services like voice and video communications, traditionally supported by circuit-switched networks, make use of IP. The move by telecom operators' corporate customers into the IP world, and the need for interoperability between private and public networks, drives the network operators to adopt IP into their backbone networks.

New technologies that are capable of transmitting IP directly over fiber using SDH framing have already emerged. Such a protocol is packet over SONET/SDH (POS), which is capable of transmitting IP packets through SDH frames in all SDH rates. In other words, gigabit IP routers could be equipped with SDH line interface cards and can be connected either directly to each other or through the usual SDH network. The overhead of POS is significantly lower for IP traffic (around 6%), but such interfaces are still very expensive. In any case, the ATM layer is missing, resulting in IP being directly transported over SDH (IP/SDH). This is an advantage in most cases, as it lowers complexity and eases network management.

One step further is the scenario of POS over DWDM, where mainly high-speed interfaces such as STM-16 or STM-64 could allow routers and switches to be connected directly to a DWDM system without going through the SDH network. So there will be cases where data and voice traffic are multiplexed by SDH before going over DWDM and cases in which IP data connects directly to DWDM without any SDH interference.

The above reasons explain why ATM is disappearing from the core of the WAN networks and being pushed back to the access part of the WAN, where it seems they will stay for some years.

1.1.3 Deployment of IP Over WDM

Although IP and WDM technologies are expected to become the dominant network technologies, they will be introduced gradually. The current infrastructure, with its mix of different types of equipment, ranging from PDH to SONET/SDH and ATM, will slowly evolve into a network in which IP traffic is transported directly over a WDM network, eliminating the need for intermediate layers.

In these evolutions, there is a factor that cannot be neglected: the majority of today's communication networks have been deployed for voice communications. These legacy networks cannot simply be discarded and replaced by new IP over WDM networks. This means that network providers will have to manage several different technologies concurrently in their networks. To provide connectivity between two points in the network across several technology domains requires manually coordinated provisioning and configuration actions on different network management systems—a tedious, time-consuming, and error-prone task. It is obvious that this situation greatly increases the complexity of network management of end-to-end connectivity.

Next generation networks built by the service providers will span multiple transport technologies, and, among others, national Internet service providers will probably choose direct IP over WDM networks.

More than one vendor will provide the equipment in these networks. This situation creates network environments where one technology is provided by one vendor, and a different technology is provided by a different vendor. Technology providers supply element and network management systems to manage their technologies, causing a creation of a *smoke stack* network management environment to the service providers.

The network management situation is further complicated by multivendor support within a single technology domain (e.g., WDM), and service provider businesses need to partition the management of their growing networks. Thus, the definition of management domains is driven by the mix of technologies, vendors, and business needs present in the given service provider environment. Lack of integration and sheer complexity of the tools themselves has become a barrier to the development of new applications, as well as the exchange and sharing of data captured by these individual network management and provisioning tools.

1.2 Structure of the Book

The main objective of this book is to present the results of about two years of research in the field of the integrated management of IP over WDM conducted under the umbrella of the WINMAN project. We considered it necessary to present these results within the perspective of the current and future challenges of the technologies that play the key roles, namely the network technologies to be managed and the management technologies themselves. This technology perspective is presented in the first part constituted by Chapters 2 to 5 as well as in a second one constituted by Chapters 12 and 13. The first part provides background while the second part depicts a framework looking at other current research projects and the most advanced foreseen scenarios. The central part of the book, constituted by Chapters 6 to 11, describes the problem, the proposed solutions, and their assessment in realistic environments.

Chapter 2 is an explanation of the rationale and market drivers for developing an interdomain network management system and the focus on IP over WDM network technologies. The "classic" way of network management will be compared with next generation views on this topic, and the expected evolution path towards the modern management paradigms will be described.

Chapters 3 and 4 present a closer look at the specifics of IP and WDM management separately. From this part, it will become clear what the current trends and common practices are, and the differences and commonalities between the management of these network technologies will become apparent.

Chapter 5 deals with the issues that arise when the IP and WDM technologies are combined and integrated into a network that is able to provide end-to-end connectivity in a highly automated fashion. Subsequently, Chapter 6 presents the concept conceived in the WINMAN project to face the challenges of multitechnology network management. This is followed by Chapter 7, which contains a detailed description of the system architecture, and Chapter 8, which details the implementation of the WINMAN solution. Finally, Chapter 9 is an overview of the software development platforms, processes and tools, and the selection criteria adopted.

Verification of the earlier mentioned management concepts has been performed on a testbed built from WDM rings and IP routing equipment, resembling as much as possible a realistic IP over WDM environment. This testbed is described in Chapter 10. The way the experiments for evaluation of the management concept were performed is also described and the results covering functional and nonfunctional aspects are presented in Chapter 11.

Because the WINMAN project is not the only activity in the area of combining network technologies, and in particular IP over WDM, Chapter 12 is devoted to an overview of one of the most representative alternatives to the WINMAN approach in the European research arena—namely, the control plane approach adopted in the Information Society Technologies (IST)–layers interworking in optical networks (LION) project.

Finally, Chapter 13 ends the book with an outlook on the expected evolution of network and management for the deployment of optical networks, with the main source being the results of the IST–(Optical Technologies in Motion for the IST) OPTIMIST project.

Appendix 1A: The Context of the WINMAN Project

Since the early 1980s, the European Commission has established programs to stimulate research and development in numerous areas. The goal of these programs was to raise the level of the European state-of-the-art technology to that of the United States and Japan. A significant area in these programs has always been telecommunications.

Within these programs, consortia of European companies and research institutes are to team up and define projects with clear objectives and well-defined results. Upon approval of a project proposal, the consortium executes the project and shares the achieved results with the European Community. The project is partially funded by the European Commission.

Starting in 1987, the first program defined by the European Commission was the Research and Advanced Communications in Europe (RACE), phases I and II (1987–1992, 1992–1995, respectively). These frameworks were followed by the Advanced Communications Technologies and Systems (ACTS, 1995–1999). The fifth framework program (FP5) is the so-called IST, which started in 1999 and will finish in 2003. The sixth framework program (FP6) was launched at the end of 2002.

The project called WINMAN (http://www.telecom.ece.ntua.gr) is a project under the umbrella of the IST program mentioned earlier. The project duration is 33 months starting on July 1, 2000.

Appendix 1B: The WINMAN Consortium

The WINMAN consortium is composed of the following companies and universities:

- Lucent Technologies Nederland B.V., Netherlands, www.lucent.com;

- Ellemedia Technologies, Greece, www.ellemedia.com;

- Telefónica Investigación y Desarollo, Spain, www.tid.es;

- Portugal Telecom Inovação, Portugal, www.ptinovacao.pt;

- Hellenic Telecom Organization (OTE) SA, Greece, www.ote.gr;

- Temagon Technology & Management Consultancy Services S.A., formerly OTE-Consulting, Greece, www.temagon.gr;

- National Technical University of Athens, Greece, www.telecom.ntua.gr;

- University College London, United Kingdom, www.ucl.ac.uk;

- Universitat Politècnica de Catalunya, Spain, www.upc.es;

- T.T.I. Team Telecom International Ltd., Israel, www.tti-telecom.com;

- Felix Telecom SRL, Romania.

2

Network and Service Evolution Scenarios

2.1 Introduction

As stated in the previous chapter, today's telecommunications market is IP driven. This means that most of the already known services like voice, data, and multimedia are being shifted from their traditional provisioning networks to IP-based networks. The same move occurs with respect to connectivity services such as virtual private networks (VPNs). The advantages derived from the convergence of the traditional telco services and computer applications in the IP world are however, putting specific requirements in the network planning as well as in its operation and management. It is necessary to understand the nature of these requirements to derive the specifications of the systems playing a role in any of the processes of the life span of the services. Voice over IP, multimedia over IP, and VPNs have then been chosen as representatives to drive further specification and design.

Although IP and the WDM technologies are believed to be dominant in the medium-term future, they will be introduced gradually in an evolutionary process starting from the present situation. Therefore, it is worth looking at the contemporary transport network, to derive evolutionary scenarios to the target goal. The current situation is characterized by the coexistence of different technologies and transport mechanisms. This heterogeneity has brought about the need for their integration, not only in the transport plane, but also in the control and management planes.

Today's networks are following an evolutionary path towards the optical transport network (OTN). This type of network can be understood as organized in four layers: the dense wavelength division multiplexing (DWDM) WAN, the DWDM metropolitan network, the access DWDM network, and the distribution/customer premises network. On the other hand, the Internet community is also proposing solutions that are sometimes different from the telecommunication carriers approaches. In this respect, it is worth mentioning two initiatives, namely Internet 2 (I2) and the so-called "next generation" Internet. These two initiatives are working towards the development and deployment of advanced network applications and technologies, accelerating the creation of tomorrow's Internet.

The objective of this chapter is to highlight the requirements set by the most common applications in the IP-oriented telecommunication market as well as the trends of the underlying transport technologies. Hence, the structure of this chapter is the following: the next section is an introduction to IP-based services focusing mainly on VPNs, voice over IP (VoIP), and multimedia over IP (MoIP) services. Special attention is devoted to the specifications in the service level agreements (SLAs) of these types of services. The next section is an overview of the current transport networks that serves as a preamble to the following one, which is devoted to analyzing the most important trends in network architectures. Here we have selected the optical transport networks, the I2, and the next generation Internet initiatives. Finally, the last section is a survey of the network management systems from a technical and a financial perspective.

2.2 Requirements of IP-Based Services

This section is intended to provide an overview of the most relevant services provided by present and near future IP networks. Understanding the services offered is fundamental in order to identify appropriate requirements for the network architectures. We have selected for this purpose VPNs, VoIP, and MoIP. These are different services that impose different requirements on the network. Most of these requirements are expressed in terms of quality of service (QoS).

Talking about QoS means dealing with how it is specified, provisioned, and assured. The specification of QoS entails giving values to system-intrinsic parameters like delay, jitter, and losses, as well as to operational parameters like time to repair and time to restore service. Different services occupy different positions in QoS planes specified by pairs of those parameters. The values taken by all of these parameters are derived from the conditions specified in the contract agreed between the customer and the service provider. This contract is called the SLA.

2.2.1 VPNs

A VPN is a communication environment where access privileges are restricted to permit peer communications only within defined community of interest. VPN products and services provide the opportunity for network service providers to secure and expand their customer base in an increasingly competitive telecommunication market. Through the VPN, the corporation can provide, but not be limited to, the following services: packet telephony services, multimedia applications (e.g., video telephony, video conferencing, and interactive learning), database access, Internet and Web access, electronic messaging, inventory management, customer service, collaboration, publishing, and electronic commerce.

A number of technologies must be integrated to provide "true" VPN solutions. They range from simple dial up to leased line for access, encryption, directory integration, and system-management techniques. The goal of a network service provider is to have a unique set of technologies integrated to offer VPN services, meeting in this way the current corporation and enterprise connectivity requirements. Figure 2.1 is a summary of these technologies grouped into the different sites involved in the VPN service.

We should note that the tremendous increase of the data traffic has significant impact to the size of VPNs in terms of bandwidth. This factor, together with the need for a simplified provisioning of VPNs and the support of on-demand bandwidth request, force network and service providers as well as manufacturers towards new solutions and approaches. Optical VPN and related services is one of them. It is expected in the near future that optical VPN services will be an important optical network service. A connection between a pair of customer ports forms the basic unit of the optical VPN services. In general, optical VPN services have some similarities with the traditional VPN services with respect to the VPN construct.

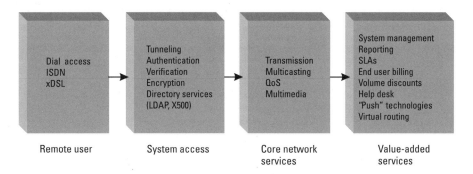

Figure 2.1 Technologies involved in offering VPN service.

2.2.1.1   Design Issues

A VPN must exhibit a set of characteristics that are embedded at its design phase and that are going to determine its management requirements. The following is a list of its most relevant characteristics:

- *Any-to-any connectivity.* The VPN must have the capabilities to provide access and communication among two or more sites. VPNs are built upon the notion of efficient and secure data tunneling between end points. By using the tunneling concept, the remote access server of the network provider wraps the user data inside packets, which are routed through the carrier's network or even across multiple networks up to the tunnel end point, where the packet is unwrapped and delivered to its destination.

- *Security.* Given a tunnel that defines a method for carrying private information across the shared network, encryption can be used to protect the contents of the tunnel. VPNs support standard forms of cryptography, including public key cryptography and data encryption standard (DES) cryptography. Compared to public key cryptography, DES is a symmetric cryptosystem. When used for communication, both the sender and the receiver must know the same secret key, which is used both to encrypt and decrypt the message.

While DES is CPU intensive, public key cryptography is 1,000 times more expensive in terms of CPU cycles. Hence, public key technology is best used as a secure alternative to password-based authentication and for key distribution. It should also be used for bulk data encryption. Due to the requirements to control and distribute keys, a significant management infrastructure is required to control the technology.

IPsec is a standards-based technology that governs security management in IP networks. Additionally, IPsec provides a standard way to exchange public cryptography keys, specify an encryption method (e.g., data encryption standards [DES or RC4]), and specify which parts of packet headers are encrypted. All of these features are extremely important, especially for emerging applications like electronic commerce, where personal data (e.g., credit cards numbers and personal identification numbers) should be transmitted in the public Internet.

- *Scalability.* Scalability must include ways to expand the capacity of the existing devices in the network. As an example, in the case of a remote site requiring more connections, another hub can be inserted within the existing architecture. Scalability allows business expansion. It eliminates forklift upgrades and usually provides load balancing and redundancy. Scalability greatly depends on the networking technology (i.e., layer 2 versus layer 3 tunneling) used to implement VPN services.

- *Network resiliency.* One of the challenges that the service providers face is the network availability. It is therefore important for the VPNs to provide efficient mechanisms for error diagnosis and recovery mechanisms.

- *Reliability and flexibility.* Reliability in a VPN means availability of services at all times, similar to the telephone network. This necessitates that redundancy features are added to allow automatic recovery of failed devices.

- *Usability.* The VPN· must be very easy to understand and use. For a VPN solution to be successful, the end users of the VPN should be able to use their services without realizing that they rely on a given VPN mechanism. For instance, the VPN must be transparent to the fact that under specific user requirements, tunnels can be established and torn down.

- *Management capabilities.* Management functionality of a VPN must cover cost-effective provisioning mechanisms, management, and billing of services with advanced monitoring and automated flow-through systems to quickly roll out new services and support SLA fulfillment.

- *QoS.* The support of QoS for VPN is not a trivial issue. The incoming traffic should be appropriately conditioned in order to fit in the agreed levels, and at the same time sophisticated traffic engineering (TE) mechanisms should be used in the core of the network to guarantee the transmission of the data according to the specified QoS agreements. In this respect, the Multiprotocol Label Switching (MPLS) framework is an ideal candidate for that purpose because it provides powerful TE capabilities as well as support for guarantee QoS.

2.2.2 Application-Oriented Services: VoIP and MoIP

During the past few years, both the traditional circuit-switched networks and the emerging IP data networks deployed throughout the world have been considered as two separate worlds [1–3]. These worlds have started to converge by examining the possibility of offering VoIP and MoIP. VoIP offers not only advanced voice services, but it also provides a means of connectivity to the existing voice networks.

The major factor driving towards this convergence is the requirement for service-driven networks. The benefit of IP is that it is a common transport for all types of networks. In the deregulated world of the near future, provision of advanced services is the major goal offered by the carriers to their subscribers.

2.2.2.1 SLA Specification for Voice and Multimedia Services

VoIP is an alternative to the traditional telephony service, and as such users expect from it a quality similar to that which they have been provided for years. Although the quality of this service, as perceived by its users, is a very subjective question, it is well known that there are a number of measurable technical parameters with a strong influence on the final quality of the conversation. ITU-T has made important efforts to define desirable performance levels on the telephony network and has published several recommendations addressing these issues.

An SLA between a customer and a VoIP provider would ideally take into account these recommendations on parameters such as end-to-end delay or echo attenuation, and it would also include quality goals usually applied to the telephony service, such as establishment time or percentage of failed calls.

A bidirectional MoIP application has similar requirements to those of VoIP, in terms of jitter and end-to-end delay. If the communication is only unidirectional, requirements are not so strict, but still jitter and delay should be bounded. In both cases (bidirectional and unidirectional), multimedia applications are more bandwidth demanding and require less packet loss than VoIP applications in order to achieve a better quality. Accordingly, the contents of an MoIP SLA will not differ much from a VoIP SLA; the differences are only the order of magnitude of the parameters.

The SLA will contain provider and also customer commitments, the ways in which all considered parameters are going to be measured, and the actions to be taken if any of the commitments are not met (i.e., penalties and discounts). The contents of the SLA may differ depending on the type of terminals used, whether traditional telephones (which need a gateway), IP telephones, or PCs (which do not need the gateway).

The most common parameters of a VoIP/MoIP SLA are:

- *End-to-end delay.* ITU-T defines a maximum acceptable end-to-end unidirectional delay in its G.114 Recommendation [4] of 150 ms, but today networks typically work with smaller values. The SLA could specify, for instance, a maximum delay of 100 ms for 99.99% of calls and an average delay of 60 ms. For unidirectional communications, this value could be greater, provided that the jitter is kept under acceptable values.

- *Jitter.* Jitter refers to a distortion of interpacket arrival times compared to the interpacket times of the original transmission. Removing jitter requires collecting packets and holding them long enough to allow the slowest packets to arrive in time to be placed in the correct sequence. This causes additional delay. The two conflicting goals of minimizing delay and removing jitter have originated various schemes to adapt the jitter buffer size to match the time-varying requirements of network jitter removal. This adaptation has the explicit goal of minimizing the size and delay of the jitter buffer, while at the same time preventing buffer underflow caused by jitter. At least, the maximum and average values of the jitter should be defined in the SLA.

- *Packet loss.* Voice/audio/video quality degrades significantly when the number of lost packets exceeds a certain percentage. This percentage depends on the voice compression algorithm (or no compression at all) in use. This degradation is aggravated by the fact that packet loss, usually due to network congestion, shows a bursty nature. So, meanwhile the average loss ratio keeps under acceptable values, the communication may get temporarily cut. The SLA should in turn define desirable packet loss values for small time intervals (e.g., 5 % during 3 seconds).

- *Bandwidth.* The minimum bandwidth that the network will support should be agreed between the provider and the customer and is to be included in the SLA. This bandwidth, together with the compression applied to the voice, audio, or video, determines the number of simultaneous sessions that can be established.

- *Rejected calls.* This parameter is commonly used in traditional telephony networks as a means to measure its correct dimensioning. For a provider to guarantee a certain maximum percentage of failed calls (e.g., 2% of all calls), the customer must promise not to exceed a certain offered traffic load and keep under what is promised (e.g., no more than 200 calls per hour).

- *Simultaneous calls.* The SLA may specify the minimum number of simultaneous telephone calls or multimedia sessions that the service provider can always accept.

- *Voice, audio, and video quality.* The quality of voice, audio, or video, when delay, jitter, and packet loss goals are met depends mainly on the audio/video codec selected. The SLA should in turn list the algorithms supported, if the user can choose among them, or the one that will be used if not. Some of the options for audio are those specified by the ITU-T: G.711 (audio coding at 64 Kbps) [5], G.722 (64, 58, and 48 Kbps) [6], G.723.1 (5.3 and 5.3 Kbps) [7], and G.729/G.729A (8 Kbps) [8]. For video, H.261 [9] and H.263 [10], along with other competing solutions, like MPEG or QuickTime, can be used. On the other hand, the user may prefer to request a subjective quality (high, medium, or low) and leave the provider the responsibility to map the request to the appropriate audio and video encoders.

- *Establishment time, also known as postdialing delay (PDD).* It is the time between the end of user or terminal equipment dialing and the reception of the appropriate network response. It should be kept to low values (for example, less than 3 seconds for 95% of calls). Two related parameters are the start dial signal delay (SDSD), the time interval between off-hook and reception of start dial signal, and the call clearing delay (CCD), the time interval between the clearance signal from the end users and the networks return to ready-to-serve state. The ITU-T addresses these issues in its E.431 Recommendation [11].

- *Echo level.* Echo is a phenomenon found in the telephone network that disturbs conversations. It is caused by imperfect decoupling when a signal traverses from a 2-wire to a 4-wire physical link. Echo becomes a significant problem when the round-trip delay is greater than 50 ms. To reduce echo, VoIP gateways must implement some means of echo cancellation. ITU-T recommendation G.126 [12] and G.131 [13] contain several considerations about listener and talker echo, showing acceptable echo attenuation level versus the total end-to-end delay. A possible value for this parameter could be an attenuation of 25 dB for 95% of calls. Echo is not a relevant issue in multimedia communications, unless a traditional telephone conversation is part of it. In this case, the same echo compensation mechanisms should be implemented.

- *Supported features.* If services such as call waiting or call forwarding are supported by the VoIP system, its conditions of use could be part of the SLA as well. The MoIP SLA can also specify additional supported features.

- *Offered traffic.* This is a compromise reached with the customer, defining how much voice/multimedia traffic it will offer the VoIP/MoIP network. If the actual traffic offered exceeds that compromised, some (or all) of the provider commitments may not apply. For traditional telephone calls, the traffic offered is usually measured in Erlangs. For data terminals, the offered traffic will equal the requested bandwidth and thus will be measured in megabytes per some time unit (e.g., a limit of 1.000 MB a day could be established).

- *Operational parameters.* Some examples are reliability, time to repair, security, and time to provision. There are no special considerations to be made for VoIP/MoIP services.

2.3 Current Transport Networks

Asynchronous transfer mode (ATM) has been the dominant transport technology for most wide area packet-switched networks for the last five years. ATM, combining transmission, switching, and multiplexing functions based on fixed-length cells, appeared a promising solution not only for core networks, but also for access networks. The characteristic of being asynchronous as opposed to the synchronous transfer mode (STM) of TDM was considered ideal for Internet-type data networks, whose traffic comes most of the times in bursts. In addition, the characteristic of being connection oriented matched the concept of traditional circuit-switched-based voice calls. Other features, such as the sophisticated traffic control and management, the support of user-defined QoS classes, and the use of optical fibers for the transportation of the cell-based traffic, presented ATM as the all-in-one network solution.

Nevertheless, ATM was not accepted in the LAN environment, where 100BaseT Ethernet was significantly cheaper, more convenient, and more importantly, supported thousands of IP applications running on top. On the other hand, neither ATM PC cards nor native ATM applications reached the real end users and were only part of research projects. The interworking of IP with ATM gave the opportunity to use the "popular" user-friendly IP applications. However, this pushed ATM to the WAN networks, most of the time not even taking into account the ATM QoS classes, but rather sticking to the usual "best-effort" IP service. The continuously growing demand for IP applications and the support for real-time and nonreal-time multimedia applications had as an immediate effect the demand for higher IP bandwidth.

As far as the STM, the SONET/SDH transmission protocol was the evolution of PDH networks based on TDM technology. All incumbent telecom operators currently operate SDH networks. SDH technology has significant advantages supporting high and variable transport rates, sophisticated management, automatic protection techniques, and easy path provisioning schemes using add/drop capabilities. SDH succeeded in satisfying the needs and growing demand for better quality mass plain old telephone system (POTS) service and integrated services digital network (ISDN) telephony. In parallel, as a result of the smooth interworking with the ATM protocol, SDH captured part of the ATM or other type data traffic into its network.

PDH, beyond the lower bit rates, lacks the significant advantages of SDH and thus suffers from the lack of network management functions, easy end- to-to-end path provisioning, and automatic recovery. This makes it expensive, inflexible, and inconvenient.

The outcome, based on the different needs regarding the services offered to the users and the corresponding traffic type, was that there was a balance mainly between the three different protocols, IP, ATM, and SDH, each one with its pros and cons according to the different service types. IP dominates the LAN area, where all of the applications feed the users. SDH is used in the WAN area and specifically in the core network to multiplex mainly traditional telephony traffic, while ATM was used in the WAN to switch data-type traffic. In most cases, ATM worked on top of a physical layer, either SDH where traffic demands were higher than the least SDH traffic rate, STM-1, or 155 Mbps, or PDH for lower traffic rate demands. Thus, the integration of multiple layers was soon a reality forming stacks such as IP over ATM over SDH (IP/ATM/SDH) or IP over ATM over PDH (IP/ATM/PDH). In other words, most wide area traffic, including IP, is converted to ATM cells and then transported over SDH paths. This usually occurs because IP traffic needs to be multiplexed with ATM traffic or other TDM traffic to make transport cost effective.

The rapid growth of the Internet market opened up many new market opportunities, and this was the main reason why ATM was squeezed between the IP and the SDH. IP is the driving force, and the demand for Internet bandwidth is increasing rapidly, reaching traffic rates of a gigabit per second for WAN environments. ATM technology reaches its limits around 2 Gbps due to the inner segmentation and reassembly (SAR) process of the ATM adaptation layer (AAL). AAL is responsible of segmenting and bringing together higher layer frames. In addition, ATM brings an extra overhead to IP being encapsulated using AAL5—the infamous cell tax, which is around 17%. In other words, the total real capacity of ATM virtual paths/channel available for IP traffic is being taxed by a significant percentage, visible especially in high transmission rates.

New technologies have already emerged that are capable of transmitting IP directly over fiber using SDH framing. Such a protocol is packet over SONET/SDH (POS), which is capable of transmitting IP packets through SDH frames in all SDH rates—STM-1, STM-4 (622 Mbps), STM-16 (2.5 Gbps), and the maximum of STM-64 (10 Gbps). In other words, gigabit IP routers could be equipped with SDH line interface cards and can be connected either directly with each other (POS over dark fiber) or through a usual SDH network (POS over SDH). The overhead of POS is significant lower for IP traffic (around 6%) compared to ATM, but such interfaces are very expensive for the time being. In any case, the ATM layer is eliminated, resulting in the direct IP over SDH (IP/SDH) approach.

One step further is the scenario of POS over DWDM, where mainly high-speed interfaces (e.g., STM-16 or STM-64) could let routers and switches connect directly into a DWDM system without going through the SDH network. So there will be cases where data and voice traffic are multiplexed by SDH before going over DWDM and cases in which IP and ATM data connect directly to DWDM without any interference from SDH. These are the main reasons why ATM is being thrown away from the core part of the WAN networks and pushed only to the access part of the WAN, where it seems that it will stay for some years. ATM capable of producing virtual paths/channels with a determined peak cell rate can serve different low broadband rates, so as to aggregate the traffic to the SDH backbone.

The move by telecom operators' corporate customers into the IP world, and the need for interoperability between private and public networks, drives the telcos to adopt IP in their core networks as a means of unifying traffic types that are compatible with their customers' networks. Gigabit Ethernet (GbE) is the new dominant fast-switching IP-enabling technology for the LAN and moreover for the MAN environment, putting aside the usual framing transmission techniques like SDH. At least for distances less than 100 km, GbE starts dominating the converged LAN and MAN data networks using only WDM or DWDM equipment as transport technology without any framing, resulting in IP over GbE over WDM (IP/GbE/WDM). GbE will reach traffic rates up to 10 Gbps, named for this reason 10-Gigabit Ethernet.

WDM technology started being used mainly in point-to-point SDH WAN links, multiplying the capacity that a single fiber can transmit by using more than one wavelength in a single fiber. DWDM steps in nicely here by lowering the cost of necessary equipment and by solving the bandwidth bottleneck reached by the TDM limits. Besides the *long-haul* WDM mentioned earlier, WDM or DWDM equipment are now being employed into *medium-haul* or metropolitan areas as well, called *metro WDM*. Metro WDM will gradually throw aside the SDH equipment, mainly in the access part, feeding the WDM equipment with traffic.

In a couple of years, WDM equipment will form the all-optical transport network in the backbone, functioning as an optical pipe multiplexing different types of signals such as GbE, ATM, and SDH. This will be the case sooner for competitive telecom operators. Incumbent telecom operators have invested significantly in SDH equipment, and there will be a longer period of time of them having both, sometimes as *overlay networks*, so that their investments are paid off. As far as the integration of IP with WDM, there will be a period of time where there is no direct interaction between the two layers, named the *overlay approach* or *model*. Integration of IP layer with WDM layer is still under research with approaches like MPLS and differentiated services (DiffServ) over DWDM and will be reported in detail hereafter.

2.4 Trends in Network Architectures

2.4.1 OTN

OTN [14] represents a natural next step in the evolution of transport networking. As an evolutionary result, optical transport networks will follow many of the same high-level architectures that were followed by SONET/SDH (i.e., optical networks remain connection-oriented multiplexed networks). The major differences derive from the form of multiplexing technology used: TDM for SONET/SDH versus wavelength division for OTNs. To satisfy the short-term need for capacity gain, large-scale deployment of WDM point-to-point line systems will continue. As the number of wavelengths grows, and as the distance between terminals grows, there will be an increasing need to add or drop wavelengths at intermediate sites. Hence, flexible, reconfigurable optical add drop multiplexers (OADMs) will become an integral part of WDM networks. As more wavelengths become deployed in carrier networks, there will be an increasing demand to manage capacity. In much the same way that digital cross connects emerged to manage capacity into the electrical layer, optical cross connects (OXCs) emerge to manage capacity at the optical layer.

Figure 2.2 depicts an OTN scenario for wide area, metropolitan, and high-capacity access networks. Initially the need for optical-layer bandwidth management will be most acute in the WAN environment. The logical mesh-based connectivity needed will be supported by means of physical topologies, including OADM-based shared protection rings and OXC-based mesh restoration architectures. As bandwidth requirements emerge for the metropolitan and access environments, the OADMs will be used there too.

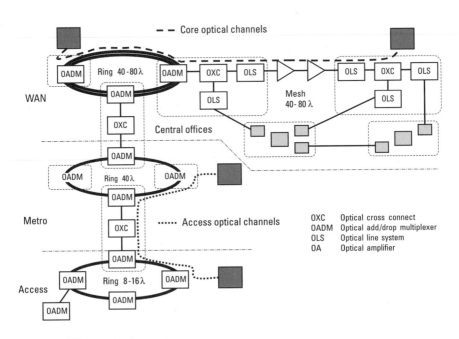

Figure 2.2 OTN architecture.

Advances in optical fibers are making the available transmission window much wider, and it is expected that fibers will be able to carry more than a thousand wavelengths in the future. Historically, the wavelength range between 1,350 and 1,450 nm has not been used because of the high fiber attenuation over much of this span caused by the presence of hydroxyl (OH-) ions. These ions are a residual impurity from the fiber manufacturing process, causing an absorption peak near 1,385 nm. Recently a breakthrough was announced in fiber for metropolitan applications, which eliminates the water absorption peak in the fiber's attenuation curve, thus making an additional window available for transmission. In effect, this makes the entire spectrum from 1,300 to 1,600 nm available for transmission. This much broader available window leads to a new concept, which we can call broad WDM (BWDM). Particularly for short-distance applications, where the need is for diverse signals at multiple bit rates, the signals can be transmitted using wavelengths that are not so closely spaced as in the ultradense WDM used in long-distance transmission. This will make it possible to cost effectively support various services such as high-performance computing, videoconferencing, broadband access (including wireless), multimedia, and the Internet.

Figure 2.3 is another view for the future network architecture, which follows the evolutionary path towards the OTN. The network is organized in four layers: the WAN backbone, the metropolitan core network, the access DWDM network, and the distribution/customer premises network.

The first layer is the WAN backbone, the so-called *long-haul* network. This layer provides the intercity DWDM or ultradense WDM backbone network and will normally connect central offices located in different cities. It consists of point-to-point DWDM connections and STM-64/STM-16 rings. Optical networking products that can be placed in this layer are optical line systems (OLS) and terabit routers/switches. Apart from DWDM links, the OLS may be connected with other cities utilizing SDH links.

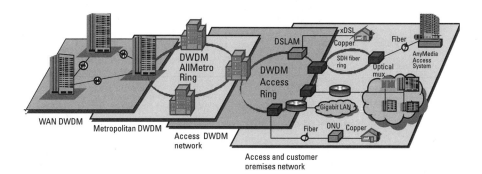

Figure 2.3 DWDM network hierarchy.

The next layer is the metropolitan DWDM network, also called the *metro* or *regional* network. The metropolitan WDM equipment will be organized in OADM-based shared protection rings, depending on the service provider's desired degree of bandwidth *overbuild* and survivability time-scale requirements. It may be connected either directly to an OLS or to a terabit router performing traffic routing at optical (wavelength) or IP layer accordingly. The metropolitan DWDM systems deliver quite a large number of wavelengths, and distribute channels of traffic to more nodes located in the city region.

The access WDM network is primarily the same as the metropolitan WDM network, but it is adapted for a smaller number of wavelengths. Its role is to distribute traffic to the access and customer premises network. Various systems can be attached to the access ring (e.g., x-type digital subscriber loop (xDSL) DSLAMs, SDH fiber rings, SDH direct links, and ONUs). In all cases, the edge device can be either connected directly to the optical terminal or a core switch can be utilized to concentrate the traffic.

For the enterprise network, many alternatives can be considered. One example is to use an ATM switch or a gigabit router as edge devices. The access interface can vary from 25 to 155 Mbps, or even to 625 Mbps. At the small office home office (SOHO) access, traffic will be normally distributed via an xDSL interface from a modem or a residential gateway.

In Chapter 12, we provide a complementary view of the OTN evolution, pointing out the deployment time scale of different technologies in the earlier mentioned network levels.

2.4.2 I2

I2 [15, 16] is a consortium being led by over 180 universities working in partnership with industry and government to develop and deploy advanced network applications and technologies, accelerating the creation of tomorrow's Internet. I2 is recreating the partnership among academia, industry, and government that fostered today's Internet in its infancy. The primary goals of I2 are to:

- Create a leading-edge network capability for the national research community;

- Enable revolutionary Internet applications;

- Ensure the rapid transfer of new network services and applications to the broader Internet community.

I2 is not a separate physical network and will not replace the Internet. I2 brings together institutions and resources from academia, industry, and government to develop new technologies and capabilities that can then be deployed in the global Internet. Close collaboration with I2 corporate members will ensure that new applications and technologies are rapidly deployed throughout the Internet. Just as e-mail and the World Wide Web are legacies of earlier investments in academic and federal research networks, the legacy of I2 will be to expand the possibilities of the broader Internet.

I2 and its members are developing and testing new technologies, such as IPv6, multicasting, and QoS that will enable revolutionary Internet applications. However, these applications require performance not possible on today's Internet. More than a faster Web or e-mail, these new technologies will enable completely new applications, such as digital libraries, virtual laboratories, distance-independent learning and teleimmersion. A primary goal of I2 is to ensure the transfer of new network technology and applications to the broader education and networking communities.

It is expected that the capabilities needed to use new technologies and applications being tested and developed by I2 and its members will be built into upcoming generations of commercial products. I2 corporate partners are working closely with the I2 community to expand the capabilities of their products and services as well as the global Internet. For example, just as most personal computers sold today include the ability to use the Internet, tomorrow's commercial products will include the ability to use advanced networking capabilities.

University research and education missions increasingly require the collaboration of personnel and hardware located at campuses throughout the country in ways not possible over today's Internet. Moreover, universities are a principal source of both the demand for advanced networking technologies and the talent needed to implement them.

Participation in I2 is open to any university that commits to providing on-campus facilities that will allow advanced applications development. The investment this requires may be more than many institutions can manage right now. However, I2 also supports collaboration by I2 universities with nonmember institutions. Fifteen years ago, connecting to the Internet could be as expensive as participating in I2 today. As the technology dropped in price, the entire academic community benefited from the efforts of the initial research participants. Deployment of I2 technology will follow a similar pattern.

A key goal of I2 is to accelerate the diffusion of advanced Internet technology, in particular into the commercial sector. I2 will benefit nonuniversity members of the educational community as well, especially schools and public libraries. I2 and its members aim to share their expertise with as wide a range of computer users as possible. This approach characterized the first Internet, and it can work again today.

2.4.3   Next Generation Internet

The *next generation Internet (NGI) initiative* [17] is a U.S. program designed to fund and coordinate federal agencies and academia to design and build the next generation of Internet services. Although the program was first proposed in 1996, it is still unclear how the NGI initiative will complement other initiatives, such as the NSF's very-high-speed backbone network service (vNBS) and I2.

The research and development needed to address this is beyond the scope of any single institution, company, or industry. The NGI initiative, with its broad agenda and ability to involve government, research institutions, and the business sector, is a timely program that will address these challenges. The goal of the NGI initiative is to conduct research and development in advanced networking technologies, to demonstrate those technologies in testbeds that are 100 to 1,000 times faster than today's Internet, and to develop and demonstrate on those testbeds revolutionary applications that meet important national needs and those that cannot be achieved with today's Internet.

The NGI initiative is expected to create the foundation for the networks of the twenty-first century, setting the stage for networks that are much more powerful and versatile than the current Internet. The NGI will also stimulate the introduction of new multimedia services in our homes, schools, and businesses as the technologies and architectures designed and developed as part of the NGI are incorporated into products and services that are subsequently made available to the general public.

The NGI initiative is part of a highly successful ongoing multiagency research and development program. It is a key component of the research and development activities of the Large Scale Networking Working Group of the Subcommittee on Computing, Information, and Communications.

The NGI initiative has three goals:

1. To advance research, development, and experimentation in the next generation of networking technologies to add functionality and improve performance.

2. To develop a NGI testbed, emphasizing end-to-end performance, to support networking research and demonstrate new networking technologies. This testbed will connect at least 100 NGI sites—universities, federal research institutions, and other research partners—at speeds 100 times faster than today's Internet and will connect on the order of 10 sites at speeds 1,000 times faster than the current Internet.

3. To develop and demonstrate revolutionary applications that meet important national goals and missions and that rely on the advances made in the first two goals. These applications are not possible on today's Internet.

The NGI initiative is closely related to I2. In fact, the NGI will rely on I2 for advanced campus-based, local area, and select regional network infrastructure. The initiative also will rely on substantial matching funds from its private sector partners.

2.5 Business Drivers for Management Systems

The following sections describe the current situation in most national incumbent operators' networks where different management systems exist for the different available technologies. In most cases, a national network consists of multitechnology, multivendor equipment and corresponding management systems capable of managing in most cases solely the equipment of a specific vendor. In the following lines, the single domain management systems are presented by briefly describing the different management layers and functionality taken from real situations.

2.5.1 SDH and ATM Management Systems

Usually the SDH or the ATM network consists of multivendor equipment. Also, different types of elements are present with different capabilities, such as terminal multiplexers, add/drop multiplexers, digital cross connects, and switches. With respect to the SDH equipment that is currently in service, we know that the growth over the last three years of the total European SDH market has reached $4 billion in 1999, $5 billion in 2000, and $5 billion in 2001.

Each vendor provides the corresponding management system, covering element management layer (EML), functionality of the telecommunication management network (TMN), and potentially also the network management layer (NML) functionality. Sometimes the element managers (EMs) provide a subnetwork view module, enabling the visualization of a whole subnetwork. In other words, it provides a partial NML view. The EMs have complete configuration management capability. Any cross connection in any of the managed types of network elements can be performed through the EMs. In addition, fault and performance management functionality is supported, mainly dealing with alarm reporting, filtering, and logging relevant to fault and performance data.

The situation and structure of the NML is similar to that of the EML. The NML is capable of having a network view for each of the vendors' equipment and able to create end-to-end paths inside the geographical area that a single vendor is covering. The network managers are able to configure and activate circuits and paths. Circuits are selected by using a variety of algorithms and rules. In order to create an end-to-end path crossing the different geographical/ multivendor areas, a special order should be delivered to each one of the management platforms. If all of the vendors provide northbound interfaces in their network management systems (NMSs), which comply with specific standards, then it might be possible to integrate all of the vendor-specific systems in one platform, but usually with mediocre results and semiautomated procedures. This is because the integrated management is not just the superposition of management layers. In other words, ensuring compatibility is not necessarily synonymous with providing the necessary functionality. With respect to network layer fault and performance management functionality, NMSs usually support additional functions, such as fault correlation, analysis, localization and recovery, and performance analysis and threshold cross reporting for end-to-end connections.

2.5.2 WDM Management

Similarities, especially to the SDH case illustrating multivendor equipment inside a national network, have been reported for WDM networks. The only difference is that the vendors have shown significant efforts in integrating the SDH management systems with the WDM management systems. As a first step, a light integration of management systems is accomplished by encapsulating the WDM management functionality in the SDH management systems. In other words, a common look and feel environment is given for both platforms under a single umbrella system. This integration can be considered high-level integration, incorporating two management systems in one platform, with common look and feel. The functionality integration (e.g., providing SDH connectivity in a mixed SDH/WDM network, creating a mixture of SDH and optical paths) is still under development in most cases. The total European WDM market was $532 million in 1999, $1.745 billion in 2000, and $2.1 billion in 2001. Obviously, less WDM is operational as compared with SDH network infrastructures. WDM management systems have yet to reach full maturity. Systems for managing WDM networks are considerably behind those for SDH. This is due to a variety of reasons; for instance, the lack of optical network protection mechanisms similar to SDH, and incorporation of the additional value (wavelength services) that WDM has to offer in a reliable network management system. It is hard to make predictions when this will catch up. In fact, with the emergence of third generation SDH the time for WDM network management to mature is expected to extend.

2.5.3 IP Management Systems

IP networks have evolved, for many years, deploying a complete set of protocols and applications suitable for delivering best-effort traffic. Management was not a priority because networks relied on very powerful control-plane protocols to reconfigure themselves in response to the addition of new elements or in case of failure. In this context, simple network management protocol (SNMP) has been developed with monitoring capabilities in mind, but with poor support for configuration management.

The previous cases, illustrating multivendor equipment inside a national network, can be the case for IP networks, especially for the incumbent operators. Nevertheless, competitive newcomer operators, which are more flexible, usually have greater homogeneity in their national networks. In addition, the SNMP protocol dominates the management of IP equipment, giving the opportunity to management platform vendors to be able to manage multivendor equipment, provided that the corresponding management information bases (MIBs) are supported or can be loaded upon demand. Even though the philosophy behind IP networks was to be connectionless, and up to now QoS is not yet supported, IP management systems provide complete monitoring of service provisioning and fault and performance data collection.

A point of inflection in the management of IP networks has occurred since the advent of the MPLS and its use as an efficient mean to achieve TE. In fact, MPLS enabled IP networks to look like connection-oriented networks. Therefore, all that has been said with respect for the management of ATM and SDH can be adopted for IP.

As a conclusion, integration is starting to appear for a single vendor among its different technology products, such as SDH/WDM integration. Nevertheless this is a high-level integration, providing only a common graphical interface for two technologies, rather than real functionality integration in terms of configuration, fault management, or performance management. The latter is either not yet performed in commercial products or is still under development.

2.5.4 The Market Perspective

The competitive environment defining the framework of current-day network providers allows customers to choose among different companies, and, because of the apparent lack of differentiation, the degree of customers' fidelity has significantly decreased.

Although quick new service development provides a way of differentiation, it is only a temporary advantage as other competitors can immediately imitate any strategy. Therefore, *front-office* activities, which have traditionally produced the majority of the perceived value for customers, while still very important in a competitive environment, are losing weight in favor of *back-office* activities.

It is in this context that network and service management, traditionally considered *back-office* activities, allows the development of skills that cannot be easily imitated and thus gives companies a way to differentiate themselves from other competitors. Having an adequate management infrastructure is of crucial importance for network and service providers in order to keep a predominant position in such a competitive environment.

This management infrastructure must cover all of the functional areas (fault, configuration, performance, security, and accounting) and address all levels of management (network-element, network, service, and business levels). Besides, networks are becoming more complicated, shifting from public switched telephone networks (PSTNs) to hybrid networks that must interwork properly. Thus, the management infrastructure must also cover all of the technologies in use, providing a common view to network and service providers.

Some of the most important areas in which management systems can be useful for network and service providers to reach and maintain an advantageous position in the telecommunication market are:

- *Revenue increase.* One of the most important problems network and service providers must face to maximize profit is revenue increase. Management systems can help to achieve this goal in several ways:

  1. *Maximizing the usage of existing products.* To do so, network and service providers can increase the average usage of the service of existing customers and acquire new customers. In order to make customers perceive a differentiation in the services offered, network operators must provide higher quality. This will result in a bigger market share. Managing service quality is therefore an essential activity for all network and service providers.

  2. *Charging higher prices.* Service providers can charge higher rates for services that require a high degree of quality. For these kinds of services, higher prices are acceptable but they must comply with SLAs. Therefore, monitoring of SLAs is required. As a consequence, service providers need tools to manage and measure network quality and performance so they can provide some evidence of the quality of service offered.

  3. *Introducing new services.* Basic services (e.g., voice traffic and bandwidth) tend to become undifferentiated standard products (commodities). Service providers need to introduce new services that yield greater added value and profit margins, as customers are willing to pay more for them. Multiservice networks that need an adequate management support these complex services. Furthermore, service development and life cycles are continuously shortening.

  4. *Integration.* Network management must integrate in a unique system the operation of networks provided by different manufacturers

and based on different cooperating technologies. In this way, network management systems offer a unique management interface that reduces operation costs by integrating in the same system the information and management facilities that would otherwise be dispersed. This makes management tasks easier and faster and allows a unified work force in which all operations staff can handle all types of network elements. At the same time, integration makes operational staff training easier and faster. Integration also permits a unique interface to service level, reducing development costs and complexity in service-level management systems.

- *Network evolution.* Network management systems must deal with networks that are increasingly complex. These networks are continuously changing, incorporating new equipment and technology domains. Therefore, network management systems must adapt easily to changes in the network and be able to incorporate new network elements.

- *Interoperability.* Another important issue that must be considered is the integration with other management systems. Consequently, network providers need management solutions that are easy to extend and offer standard interfaces to network elements and other management systems.

- *Standardization.* Standardization has become very important in the telecommunications market, as it provides a common framework for network and service management. Network and service operators' interest in the development of management standards stems from the fact that the adoption of standards by the telecommunications industry makes the integration of network elements and management systems much easier. Thus, the use of standard solutions results in lower costs and more efficient management solutions. Therefore, network management systems should be based on standard solutions.

2.6 Summary

Telecom operators, both incumbents and new entrants, are giving significant importance to the deployment of IP networks, recognizing the trend that yields delivery of conventional and new services through the IP protocol. VPNs, voice, and multimedia over IP are representative scenarios. Nevertheless, the success of these services depends to a great extent on the availability of solutions to solve the different technological challenges. In fact, VPNs over IP must be designed to exhibit similar characteristics as conventional VPNs in terms of security, scalability, network resilience, flexibility, usability, and especially manageability and availability of QoS. At the same time, the IP-oriented application services like voice over IP or multimedia over IP cannot be regarded as best-effort services. Technologies and frameworks like MPLS, DiffServ, and IntServ must be used to provide to customers guaranteed QoS. Additionally, SLAs specifying among others the functional and operational characteristics for the provided services are used, leading to fierce competition between service and network providers in the race for acquiring the biggest share of the market.

On the side of the transport network technology, ATM has been the dominant transport technology for most wide area packet networks. This was not the case in LAN environments, where the predominant technology has been xBaseT Ethernet. On the other hand, the SONET/SDH protocol was the evolution of PDH networks and fulfilled the needs and growing demand of mass plain telephony services. SONET/SDH can smoothly interwork with ATM, and the same occurs with IP and ATM. Therefore, the integration of multiple protocol layers like IP/ATM/SDH or IP/SDH (POS) has been a reality.

The continuous increase of the traffic volume has led to fiber exhaustion problems. New methods and technologies have been explored as a way to exploit existing infrastructure, avoiding the need for new fiber installations that increase the cost of network deployment significantly. WDM is one of the technologies that addresses the problem of fiber exhaustion by multiplexing in each fiber many wavelengths (lambdas), multiplying the capacity of existing fiber. This makes the WDM technology a major, if not the dominant, candidate for future optical networks. On the other hand, the inherent limitation of ATM around 2 Gbps has also fostered the evolution to new protocol stacks like IP/SDH/WDM or IP/WDM. In the last case, there will be still some time where there is no direct interaction between the two layers (overlay model).

The OTN seems to be the logical next step in the evolution of transport networks. This network will adopt the IP, WDM, or DWDM technologies architectured around four identifiable layers—the long-haul DWDM network, the metropolitan DWDM network, the access DWDM network, and the customer premises network (this one with different coexisting access technologies).

Although the IP and WDM technologies are believed to be dominant in the future, deployed networks are not going to change radically, and they will be introduced gradually following an evolutionary path. Such evolution scenarios will be determined by the advances in the already mentioned OTNs as well as by initiatives like the NGI and I2. The NGI initiative is a U.S. program involving federal agencies and academia intended for the research and development of new networking technologies exhibiting improved performance to create a new Internet testbed and to develop and demonstrate new applications not possible in today's Internet. The I2 is a worldwide consortium with similar objectives.

The transition period from the current services and network technologies will be characterized by the complexity of the network architecture as well as the heterogeneity of network elements. The key that will diversify network and service operators will be their ability to provide in an automatic and flexible way existing and new services, making the best use of their network infrastructure. Single technology network management system partly covers this functionality, whereas new integrated network management systems for single vendor elements have already emerged. However, real integrated management systems capable of managing multitechnology, multivendor network elements are still under development. In this context, service and network integrated management systems will be the key for service and network providers in order to help them to increase their revenues, by maximizing the usage of resources, allowing them to charge higher prices due to the corresponding service differentiation, and facilitating the introduction of new services.

References

[1]   Hamdi, M., et al., "Voice Service Interworking for PSTN and IP Networks," *IEEE Communications Magazine,* Vol. 37, No. 5, May 1999, pp. 104–111.

[2]   Prycker, M. D., and T. Van Landegem, "New Network Architectures," *Alcatel Telecommunications Review,* 2nd Quarter, 1999.

[3]   Internet2, www.internet2.edu, December 2002.

[4]   ITU-T Recommendation G.114, "One-Way Transmission Time," May 2000.

[5]   ITU-T Recommendation G.711, "Pulse Code Modulation (PCM) of Voice Frequencies," November 1988.

[6]   ITU-T Recommendation G. 722, "7 kHz Audio-Coding Within 64 Kbps," November 1988.

[7]   ITU-T Recommendation G.723.1, "Dual Rate Speech Coder for Multimedia Communications Transmitting at 5.3 and 6.3 Kbps," March 1996.

[8]   ITU-T Recommendation G.729, "C Source Code and Test Vectors for Implementation Verification of the G.729 8 Kbps CS-ACELP Speech Coder," March 1996.

[9]   ITU-T Recommendation H.261, "Video Codec for Audiovisual Services at p x 64 Kbps," March 1993.

[10] ITU-T Recommendation H.263, "Video Coding for Low Bit Rate Communication," February 1998.

[11] ITU-T Recommendation E.431, "Service Quality Assessment for Connection Set-Up and Release Delays," June 1992.

[12] ITU-T Recommendation G.126, "Listener Echo in Telephone Networks," March 1993.

[13] ITU-T Recommendation G.131, "Control of Talker Echo," August 1996.

[14] ITU-T Recommendation G.872, "Architecture of Optical Transport Networks," November 2001.

[15] Archives for Internet2-NEWS, http://archives.internet2.edu/guest/archives/I2NEWS, December 2002.

[16] "Theme Features: Converging on Internet Telephony," *IEEE Internet Computing*, Vol. 3, No. 3, May/June 1999, pp. 40–91.

[17] Center for Next Generation Internet, http://www.ngi.org, December 2002.

3

Management of the IP Network Layer

3.1 Introduction

The first and second generations of the Internet have been defined to provide connectivity in the scientific community and to support the concept of the World Wide Web, respectively. For both generations, the services provided are best effort, in the sense that no guaranties of QoS are supported. In this context, the role of network management is downplayed. In fact, in terms of the widely accepted five management functional areas—fault, configuration, accounting, performance, and security (FCAPS)—and the four levels of the TMN reference model (network element, network, service, and business), we can say that management in the early and mostly in the current Internet is focused only on configuration and performance at the network element level. Moreover, configuration management is essentially a manual process that requires skills at the network-device level.

The third generation of the Internet, whose first steps are already being taken, is essentially QoS enabled. As we pointed out in Chapter 2, services like VPNs, voice, or multimedia need to be provided with strict QoS specifications that are contained in SLAs. Moreover, these services need to be customized to particular user needs, and they need to be deployed quickly and efficiently. This creates a new context where management systems, particularly the management systems at the IP layer, are the key to facing the new challenges. Therefore, we can expect that the management of this layer is extended to the five FCAPS and to the four TMN levels in a way that will require increasingly less human intervention and skills.

This chapter is devoted to highlighting the aspects of IP management in the context of the earlier mentioned evolutionary scenario, specifically addressing the most relevant scenarios to understand the solutions adopted in Chapters 6 through 11 of this book. It deals with "traditional" IP management, although it is mainly focused on the management aspects of the MPLS connectivity network model. At the same time, the chapter is heavily based on standards and frameworks recognized worldwide, such as the Internet Engineering Task Force (IETF), OSI, ITU-TMN, and the TeleManagement Forum (TMF). The promissory aspects of policy-based network management for abstraction of the implementation details are remarked upon when in favor of a high-level view of network services. Such an abstraction is a basic requirement for the dynamic provision of IP connectivity services. Some useful tools and the trends in the IP/MPLS world, characterized by the current activities in relevant fora, are also presented.

The chapter is structured in five specific sections besides this introduction and the summary. In Section 3.2, we describe the fundamentals of IP MPLS-enabled networks and applications as, although they do not belong to the IP protocol stack, MPLS is often linked to them in the QoS-enabled Internet and, in addition, they have been adopted in the context of the solutions provided later in this book. Section 3.3 is focused in the management aspects of the IP layer, starting with the case of best-effort networks and following with IP MPLS-enabled networks. Section 3.4 is entirely devoted to presenting the policy-based network management paradigm. Section 3.5 follows with an introduction of a set of management tools structured around the scope of the use of these tools. Finally, Section 3.6 addresses the trends in the IP MPLS-enabled network.

3.2 The IP/MPLS Network Model

Traditional IP networks, such as the current Internet, provide best-effort connectionless service based upon IP datagrams (packets) that traverse the network from source to destination based on a hop-by-hop routing. Each packet in a flow is an individual entity routed as such. So the whole concept of flow is nonexistent; each packet in a stream may use a different path or be lost on the way. In this context, there is not one IP *network model*, but individual managed elements (the routers or nodes) that contain several managed objects defined by MIBs. The fundamental objects in a router are its network interfaces, which provide connectivity to other routers and end systems. Furthermore, the interfaces are the most useful targets to measure usage and other performance parameters (e.g., link utilization). Nevertheless, no network model exists for the conventional best-effort Internet [1].

A new network model based on end-to-end IP connectivity has emerged as a result of the deployment of MPLS. MPLS is the latest step in the evolution of routing/forwarding technology for the core of the Internet, following many technological advances designed to support ISPs as they try to keep a step ahead of the Internet's explosive growth. Such advances include Internet backbone routers, new queuing and scheduling algorithms, IPSec, Web-caching services, directory services, and integrated routing/forwarding solutions.

MPLS delivers a solution that seamlessly integrates the control of IP routing with the simplicity of layer 2 switching. Furthermore, MPLS enables the deployment of advanced routing services because it solves a number of complex problems:

- MPLS addresses the scalability issues associated with the currently deployed IP-over-ATM overlay model.
- MPLS significantly reduces the complexity of network operation.
- MPLS facilitates the delivery of new routing capabilities that enhance conventional IP routing techniques.
- MPLS offers a standards-based solution that promotes multivendor interoperability.

The fundamental ideas of the MPLS solution are:

- Separation of the control and forwarding components;
- A label-swapping forwarding algorithm.

Figure 3.1 presents the basic operation of the IP-MPLS–based transport mechanism. A comprehensive description of the MPLS architecture, including these basic characteristics, can be found in [2], whereas the key aspects of MPLS operation follow hereafter.

In conventional IP, routers make independent decisions to send arriving packets to their next router destinations. This is done by checking the IP destination address prefix against the local IP routing table. In contrast, in MPLS,

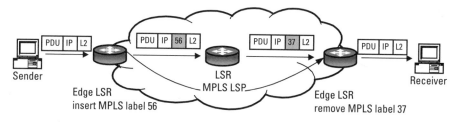

Figure 3.1 Principles of the MPLS operation.

the mapping between the packet headers and the path that will be followed up to the final end is done just once, when the packet enters into the network. The path to which a given packet is assigned is encoded into a label that is virtually transported with it as it traverses the network. As a consequence, there is no longer the need to go up to the network packet headers to decide where to send a packet arriving at a router. Instead the label is checked against a routing table that will specify the next hop in the path, similar to the way that ATM routing tables deal with virtual paths. The routers entrusted to insert and remove the MPLS labels are the edge routers of the MPLS network cloud; the routers that are able to take decisions based on such MPLS labels are called label switching routers (LSRs). The path that is assigned to a packet to cross the network is called a label-switched path (LSP). Moreover, at the path establishment the routers inside the MPLS network cloud need to inform themselves about the mapping between the labels and the LSP. This is carried out through a protocol called label distribution protocol (LDP). More detailed information can be found in [3] and [4].

Therefore, the network is characterized by end-to-end LSPs with QoS constraints that support specific traffic flows defined by forwarding equivalence classes (FECs). These FECs are classifiers based on several characteristics, such as origin and destination IP addresses or class of service (CoS) identifiers, which allow the mapping of the ingress traffic to specific LSPs. These LSPs use constraint-based routing, which can be statically defined as explicit routes by the configuration management application (a hop-by-hop list) or calculated by the routing protocol, like open shortest path first (OSPF) or intermediate system-intermediate system (IS-IS), with TE extensions. This means that the routing calculation for an MPLS-enabled network can be done off-line by a management application and then downloaded to the network devices, with possible rerouting triggered by the control plane in case of failures. The connectivity matrix can also be built by the network itself using the aforementioned TE-enabled routing protocols and the TE signaling mechanisms. In either case, this model permits the compilation of an end-to-end traffic matrix that enables the service provider to keep track of network usage on a flow-by-flow or client-by-client basis (or any other chosen criteria). This is a great aid for the network management, especially for monitoring SLAs.

3.2.1 MPLS-Based Applications

Currently there are three main applications making use of MPLS in the core of large ISP networks:

1. Traffic engineering implementation;
2. Differentiated services support;

3. VPNs deployment.

3.2.1.1 MPLS Traffic Engineering Implementation

Traffic engineering is defined as that aspect of network engineering concerned with the performance optimization of traffic handling in operational networks, where the main focus of the optimization is to minimize overutilization of capacity when other capacity is available in the network. Traffic engineering entails the aspect of network engineering concerned with the design, provisioning, and tuning of operational networks. It applies business goals, technology and scientific principles to the measurement, modeling, characterization and control of traffic, and the application of such knowledge and techniques to achieve specific service and performance objectives, including the reliable and timely transport of traffic through the network, the efficient utilization of network resources, and the planning of network capacity [5].

TE enables the mapping of traffic flows onto an existing physical topology. It provides a means of controlling the distribution of traffic across all network links so that they are more evenly utilized. The network planner is allowed to move traffic flows away from the shortest path calculated by the routing protocol and onto potentially less congested physical paths. TE is currently the primary application for MPLS because of the unprecedented growth in demand for network resources, the mission-critical nature of IP applications, and the increasingly competitive nature of the service provider marketplace. A successful TE solution can balance a network's aggregate traffic load on the various links, routers, and switches in the network so that none of its individual components is overutilized or underutilized. This results in a network that is more efficiently operated and provides more predictable service.

MPLS is well suited to enable TE in large ISP networks for the following reasons:

- Support for explicit paths allows network administrators to specify the exact physical path that an LSP takes across the service provider's network.

- Per-LSP statistics can be used as input to network planning and analysis tools to identify bottlenecks and trunk utilization, and to plan for future expansion.

- Constraint-based routing provides enhanced capabilities that allow an LSP to meet specific performance requirements before it is established.

Traffic-engineered LSPs are often referred in the literature as MPLS-TE tunnels.

3.2.1.2 Differentiated Services Support

MPLS is designed to support DiffServ [6] by means of the CoS bits in the packet header. The DiffServ model defines a variety of mechanisms for classifying traffic into a small number of service classes. To meet customer requirements, ISPs need to adopt not only TE but also traffic classification techniques. An ISP can take two approaches to support MPLS-based CoS forwarding:

1. Traffic from each LSP can be queued for forwarding based on the setting of the CoS bits carried in the MPLS header. This is like ordinary IP DiffServ applied to each LSP. All of the traffic of a given flow is carried end to end by the established LSP, reordering packets in accordance with the CoS.

2. Provision multiple LSPs between each pair of edge LSRs. Each LSP can be traffic engineered to provide different performance and bandwidth guarantees. The head end LSR could place high-priority traffic in one LSP, medium-priority traffic in another LSP, best-effort traffic in a third LSP, and so on. This way there will be different *virtual networks* for each type of traffic. The traffic from a given flow will be reclassified in the MPLS network, using different paths in accordance with the CoS.

3.2.1.3 VPN Deployment

A VPN behaves like a private WAN over the public network facilities (e.g., the Internet). To offer a viable VPN service for its customers, an ISP must solve the problems of data privacy and support the use of nonunique, private IP addresses within a VPN. MPLS provides a simple and efficient solution to both of these challenges because it makes forwarding decisions based on the value of the label, not the destination address in the packet header.

VPNs are typically constructed using four fundamental building blocks:

1. Firewalls to protect each customer site and provide a secure interface to the Internet;

2. Authentication servers to verify that each customer site exchanges data with only validated remote sites;

3. Encryption to protect data from examination or manipulation as it is transported across the Internet;

4. Tunneling encapsulation to provide a multiprotocol transport service and enable the use of the private IP address space within a VPN.

MPLS provides a simple, flexible, and powerful tunneling mechanism. An ISP can deploy a VPN by provisioning a set of LSPs to provide connectivity among the different sites in the VPN. Each VPN site then advertises to the ISP a set of prefixes that are reachable within the local site. The ISP's routing system distributes this information by piggybacking labels in routing protocol updates or by using a label distribution protocol. VPN identifiers allow a single routing system to support multiple VPNs whose internal address spaces overlap with each other. Finally, each ingress LSR places traffic into LSPs based on a combination of a packet's destination address and VPN membership information.

The most widespread model for the implementation of VPNs over IP/MPLS is the BGP/MPLS model.

3.3 Management of IP Networks

Traditional IP network operation and management has been focused mainly on monitoring faults and performance variables of the IP network elements by means of SNMP [7]. Configuration management has been left behind and is mainly performed by cumbersome vendor proprietary mechanisms, like command line interface (CLI) commands and/or nonstandard SNMP MIBs, accessible by nonstandard management software. Management engineering has a long tradition of script-based configuration management and on-site configuration of devices.

Moreover, vendors' implementation of standard SNMP MIBs are often incomplete, and it is common to find that variables that are supposed to be writeable are actually only readable, and many do not exist at all. Network management planning in such an obscure environment has always been a challenge circumvented mainly by human effort.

The continuous addition of new features to IP equipment, promoted by vendors' need to increase their market share, imposes additional limitations in the SNMP configuration management area. This fact also affects another important aspect in this area: the software maintenance of devices and operating system versions. Some network devices can read its configuration at startup from a network file server by means of the trivial file transfer protocol (TFTP) or a similar one. Changes to the configuration when the equipment is operational can be done the same way, but there are no standards for such operations. This situation regarding configuration management is still valid nowadays, beside some standardization efforts that are being undertaken (see Section 3.6).

3.3.1 Performance Management

The main reference for SNMP-based performance management of IP network elements is the MIB-II, defined by RFC 1213 and updated by RFC 2011, RFC

2012, and RFC2013. Relevant IP management subtree objects are shown in Figure 3.2 (partial view).

The most useful MIB-II groups for an SNMP management application are the `interfaces group` and the `IP group`.

The interfaces group contains information about the device's interfaces, including configuration information and statistics. Several counters can be used by a performance management application to detect usage and reliability (indicated by the error counters).

The IP group relevant information is some basic counters of traffic flow at the IP layer and the routing tables, contained in the `ipRouteTable`. Routing protocols and tables can be configured and monitored using some of these variables.

3.3.2 Fault Management

Fault management is based on traps sent from the agent to the management platform. A trap holds information of a specific structure defined by RFC 1215. Because traps are asynchronous messages signaling an event that may require attention, there are no responses from the management station to them.

Due to its initial requirement to be protocol-independent, the Internet-standard SMI does not provide a means for defining traps. Instead, the SNMP defines a few standardized traps and provides a means for management enterprises to transmit enterprise-specific traps.

3.3.3 SNMP Management of IP/MPLS Network Connectivity

As stated earlier, MPLS-enabled networks permit the consideration of end- to-end LSPs as atomic and manageable elements, though evolving from an element-centric to a connectivity-centric management paradigm.

Several existing MIBS can be used for MPLS management, in addition to some specific MIBs for MPLS management defined by IETF [8]. The MPLS MIB subtree structure is shown in Figure 3.3 [please note that the root object `mplsMIB` and other objects has no OID assignment yet by the Internet Assigned Numbers Authority (IANA)].

The MPLS-TC-MIB (IETF RFC 2579) describes textual conventions and object identities that may be common to MPLS-related MIBs.

The MPLS-LDP-MIB (IETF RFC 3036) describes managed objects used to model and manage the MPLS LDP.

The MPLS-FTN-MIB describes managed objects that are used to model and manage the MPLS FEC-to–next hop label forwarding entry (NHLFE) mappings, which take place at any ingress LSR. LSRs push, swap, and pop

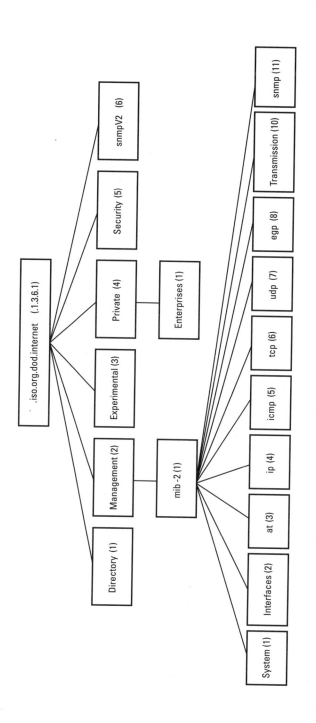

Figure 3.2 MIB II subtree.

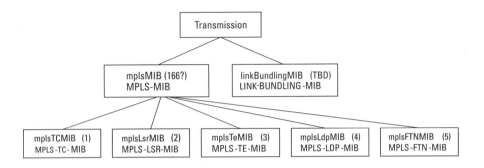

Figure 3.3 MPLS MIBs under the transmission MIB II node.

MPLS headers according to rules contained in a forwarding table called *forwarding information base* (FIB), unique for each MPLS router. The FIB can contain three different types of entries:

1. A NHLFE that contains the information necessary to forward a packet with an already assigned label; the NHLFE contains two pieces of information: the packet's next hop address, and also whether the MPLS header of the packet must be swapped or popped. If the MPLS header of the packet must be swapped, then the NHLFE also contains the new label of the packet.

2. The incoming label map (ILM), which contains the mappings between labels carried by incoming packets and NHLFE entries;

3. The FEC-to-NHLFE (FTN), which contains the mappings between incoming packet FECs and NHLFE entries.

The MPLS-LINK-BUNDLING-MIB describes managed objects that are used to model and manage the MPLS TE interfaces, as well as the link bundling relationships that may exist between those interfaces.

From a functional point of view, the most useful MIBs for management applications are the MPLS-LSR-MIB and the MPLS-TE-MIB. Hereafter we describe some of this management functionality grouped in the areas of configuration/performance and fault management.

3.3.3.1   Configuration and Performance Management

The following MPLS-LSR-MIB tables are worthy to mention:

- The interface configuration table (mplsInterfaceConfTable), which is used for enabling the MPLS protocol on MPLS-capable interfaces;

- The in-segment (mplsInSegmentTable) and out-segment (mplsOut-SegmentTable) tables, which are used for configuring LSP segments at an LSR;

- The cross-connect table (mplsXCTable), which is used to associate input and output segments together, in order to form a cross connect;

- The label stack table (mplsLabelStackTable), which is used for specifying label stack operations;

- The traffic parameter table (mplsTrafficParamTable), which is used for specifying LSP-related traffic parameters;

- Moreover, the MPLS in-segment and out-segment performance tables, mplsInSegmentPerfTable and mplsOutSegmentPerfTable, contain the objects necessary to measure the performance of LSPs, and the mplsInterfacePerfTable has objects to measure MPLS performance on a per-interface basis.

The MPLS-TE-MIB is based around a table that represents TE tunnels that either originate at a given LSR or traverse or terminate on the LSR. The following MPLS-TE-MIB tables can be used:

- Tunnel table (mplsTunnelTable) for setting up MPLS tunnels;

- Resource table (mplsTunnelResourceTable) for setting up the tunnel resources;

- Tunnel specified, actual, and computed hop tables (mplsTunnelHopTable, mplsTunnelARHopTable, and mplsTunnelCHopTable) for strict and loose source routed MPLS tunnel hops;

- Constraint-based LDP (CR-LDP) resource table (mplsTunnelCRLDPResTable) for specifying resource objects applicable to tunnels signaled using CR-LDP.

3.3.3.2 Fault Management

The four MPLS-TE-MIB notifications, namely, mplsTunnelUp, mplsTunnelDown, mplsTunnelRerouted, and mplsTunnelReoptimized, give asynchronous information about the status of the MPLS tunnels.

In case of a lack of the previously mentioned MPLS native MIBS, the MIB-II objects linked to the *mplsTunnel* (150) at the *interfaces* group can be used for MPLS monitoring. Also the conventional *linkDown* and *linkUp* traps can be used for monitoring LSPs.

3.3.3.3 Example of SNMP-Based MPLS Management: A Generic Object-Oriented MPLS-TE Application Programming Interface

The example refers to an implementation of MPLS-TE management using SNMP and Java language [9]. The application programming interface (API) uses a generic SNMP Java implementation and builds its functionality from an object-oriented information model of the devices' configuration (depicted in Figure 3.4). Even though the actual implementation is being used with Cisco routers, it is generic enough to accommodate other vendors' equipment because the specific parts are template based. Minimum changes are needed to work with different equipment. The MPLS-TE API is able to configure MPLS LSPs with all the functionality provided by the vendors' equipment.

The configuration class abstracts the configuration information (a text file in the case of Cisco routers). This class has several attributes of classes that provide further detail about the configuration; in particular the tunnel class contains the attributes for a given MPLS tunnel configured in a device. The configuration class is generic, not tailored to any vendor. A specialization is needed when parsing the configuration file and writing configuration information to the device. A template with the appropriate vendor's syntax is used to generate the configuration file that is downloaded to the device by TFTP.

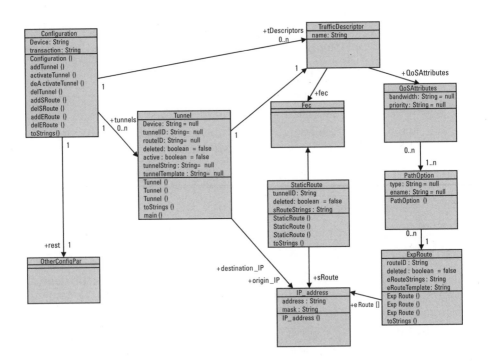

Figure 3.4 MPLS-TE API model.

Another remarkable characteristic of the API is the implementation of the concept of transactions: a configuration change is done completely or it is not done at all. We achieve this goal by transferring the complete configuration file to the device and replacing the active configuration by this new file. The whole process is controlled using a proprietary Cisco MIB for device configuration [10, 11].

This API is extensible to every aspect of the router configuration. This can be achieved by extension of the configuration class and the corresponding templates. A multivendor implementation would need specific templates for each vendor. The method for extending the API is to specialize the class `OtherConfigPar`, which at the moment contains everything but the MPLS-TE tunnels configuration.

The API supports the whole range of SNMP traps and can capture and parse them in order to notify the exception condition to any management system using the API.

Performance management functionality is currently under development and is based on standard SNMP meters and the proprietary CISCO-RTTMON-MIB [12] that enables the measurement of one-way delay (synchronization needed), round trip delay, and jitter. More information about this MIB usage can be read at [13].

3.3.4 Management of MPLS VPNs

Based on the successful deployment of the previously mentioned BGP/MPLS model for IP VPNs, many vendors are providing management software for this application. These applications are inherently multivendor because the model is a standard implemented by several IP equipment providers. These tools are mainly centered on performance and configuration management, and they enable service providers to deploy VPN services on their MPLS-enabled backbones.

Examples of these tools can be found in the Web sites of major vendors like Cisco [14] and Juniper [15], and some software companies like Orchestream [16] and WANDL [17], among others.

Also, a MIB for modeling MPLS/border gateway protocol (BGP) VPN management has been published by the IETF [18].

3.4 Policy-Based Management

The aim of policy based network management (PBNM) is to make reality a vertical integration among network-element management, network management, and application management. In fact, PBNM provides a way to abstract the service characteristics from the implementation details, allowing high-level definition of services, which are translated to configuration parameters at the appropriate management layer. For example, a VPN can be implemented using the MPLS/BGP model, the IP secure protocol (IPSec) model, the virtual routers model, or a combination of these. These details can be hidden to the service-level management and be implemented to effectively accommodate network resources. MPLS-based connectivity provisioning can also benefit from this approach, allowing network managers to think in terms of access points and traffic parameters, letting the lower level application translate these definitions into MPLS-TE tunnel definitions.

There are many standardization activities for policy management in the IETF, mainly focused on QoS issues. The IETF's RAP working group [19] has developed the common open policy server (COPS) [20] protocol to manage policy information, and the Policy Framework working group is defining a common framework for policy management, information model, and specific schema. The information model is the policy core information model (PCIM) and extensions [21], based on the common information model (CIM) [22] and the proposed directory schema uses lightweighted directory access protocol (LDAP) and directory enabled networks (DEN).

The IETF Policy work group has developed a general model for representing and managing policies. A policy is considered to be a set of policy rules, and each policy rule is composed of a set of conditions and a set of corresponding actions. All policy rules have the format "if a condition occurs, then perform an action." The main entities comprising the policy framework can be seen in Figure 3.5.

The policy decision point (PDP), as its name implies, is the point at which the policy decisions are made. On the other hand, the policy enforcement point (PEP) is the point at which the policies are enforced (usually located in the network element that supports policy enforcement). For a more detailed description of this framework, see [19].

3.5 IP/MPLS Management Tools

In this section we review many well-known tools from the Internet community plus some suggested tools for policy and MPLS management. We have classified them under the following categories: SNMP-based tools are "everyday" tools used by network managers to quickly check out the status of some MIB variable. These tools are often used in shell scripts or integrated into programs to build simple but powerful management applications. SNMP APIs are more complete tools tailored for management application programmers in order to build higher level management applications at the element-network and network-

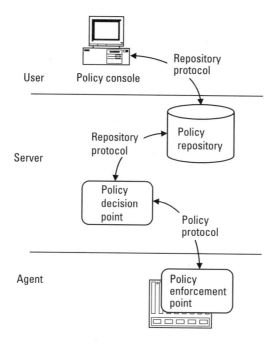

Figure 3.5 General architecture of a policy-based management system.

management layers. PBNM tools are becoming more popular, but are still experimental or limited in scope in the case of commercial tools. MPLS tools are more specific and not yet integrated into management frameworks, but allow the implementation of simple monitoring applications.

3.5.1 SNMP-Based Tools

As mentioned earlier, there is a long tradition of script-based management among the IP community. Many open source tools and SNMP APIs permit one to build basic management applications. We mention Multirouter Traffic grapher (MRTG), a graphical tool that easily permits the building of a traffic monitoring Web site. Also Net-snmp and Scotty are very useful command-line utilities and daemons, also scriptable, which perform standard SNMP queries and trap listeners.

3.5.1.1 MRTG

MRTG [23] is a tool to monitor the traffic load on network links. MRTG generates hypertext markup language (HTML) pages containing portable network graphics (PNG) images, which provide a live visual representation of this traffic, as shown in Figure 3.6. It is based in SNMP libraries of the Perl scripting language.

3.5.1.2 net-snmp

This package, based on work at the Carnegie Mellon University and the University of California, provides various tools relating to SNMP, including an extensible agent, a SNMP library, tools to request or set information from SNMP agents (e.g., snmpset and snmpget), tools to generate and handle SNMP traps, and a Perl-based MIB browser. The project is alive and new releases are launched regularly.

3.5.1.3 Scotty

This package was developed at the University of Twente and the Technical University of Braunschweig, allowing the implementation of site-specific network-management software using high-level, string-based APIs. The software is based on the tool command language, which simplifies the development of portable network-management scripts. The Scotty source distribution includes two major components. The first one is the Tnm Tcl extension, which provides access to network-management information sources. The second component is the *Tkined* network editor, a Tcl extensible graphic editor.

3.5.2 SNMP APIs

There are many freely available SNMP APIs for different languages (Java, C, Perl, and Python). An extensive reference can be found at [24]. These APIs allow the programmer to build SNMP applications, skipping the burden of building the SNMP protocol from scratch.

3.5.3 PBNM Tools

Many commercial PBNM available nowadays share some common features (e.g., they are network element managers for QoS characteristics, such as

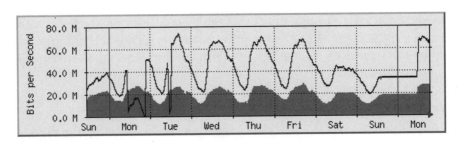

Figure 3.6 MRTG monitoring graphical output.

queuing or PHB configuration, and they usually do not offer an open interface for external applications, such as an external network level manager).

3.5.3.1 Jasmin

The Distributed Management working group (DISMAN) of the IETF has developed a standard for distributing and invoking network management scripts [25].

Jasmin is a joint project between the Technical University of Braunschweig and NEC C&C Research Laboratories to evaluate and enhance this standard by providing and studying an implementation with distributed network management applications.

The implementation supports multiple languages and runtime systems. A script MIB extensibility protocol (SMX) allows for plugging to additional runtime systems. By now, two runtime systems have been implemented in this project: one is based on the Java virtual machine and another one is support Tcl. Furthermore, the script MIB can be used for policy-based configuration management.

3.5.3.2 Ponder

Ponder is a language for specifying QoS and security policies for distributed systems. It has been developed as part of an ongoing research into the use of policies in distributed systems management [26]. Also a complete toolkit has been developed to support the users of the language. Available components include:

- *A Ponder compiler.* The compiler for the Ponder specification language, which consists of a syntax analyzer, a two-pass semantic analyzer, and a default Java code generator for policies;

- *A Ponder policy editor.* A customizable text editor for the Ponder language, written in Java, which has all the basic features of a text editor and includes features that make text editing Ponder policies easy;

- *Ponder management toolkit.* The management toolkit has been designed to allow for the addition of tools to be managed from a central management console.

The Ponder framework uses an LDAP repository for storage of managed objects and system policies. The research community has been involved in the area of policies for management of networks, identifying different types of policies [27, 28].

3.5.4 MPLS Tools

Apart from these tools, the IETF working groups are proposing some MPLS-specific tools for monitoring and carrying out basic operations and maintenance (OAM) of MPLS-enabled networks. Most proposed diagnostic tools rely upon IP (e.g., the LSP ping [29] and the generic tunnel tracing [30]).

3.5.4.1 LSP Ping

LSP ping is a simple and efficient mechanism to detect or diagnose data plane failures in MPLS LSPs, specific to reservation protocol with traffic engineering extensions (RSVP-TE) networks.

Ingress LSR sends ICMP echo_request messages over the LSP, which should be returned by the egress LSR. If after a defined time period ICMP messages are not returned, then ingress LSR sends a LSP-ping probe packet (RSVP-TE extension) to the egress LSR over the LSP under test. Egress LSR acknowledges receipt of the probe packet by returning the LSP_ECHO object in the RSVP Resv message back to the ingress LSR.

If the ingress LSR does not receive a Resv message from the egress LSR that consists of an LSP_ECHO object within a period of time, it declares the LSP as *down*. At this point, the ingress LSR should apply the necessary procedures to fix the LSP. This may include generating a message to a network-management console, tearing down and rebuilding the LSP, and rerouting user traffic to a backup LSP.

3.5.4.2 Generic Tunnel Tracing

This tool helps to reduce the time of detecting or diagnosing failures within a tunneled network. It solves some missing functionality in traditional *traceroute* command. Its main features are:

- Trace path through generic tunnels (e.g., MPLS, IP-in-IP);

- Support heterogeneous nested tunnels (e.g., IP-in-IP over MPLS);

- Trace through forwarding plane or control plane.

This tool requires a protocol that supports these applications. This protocol reveals the path between two points in an IP network. When access policy allows it, the protocol also reveals tunnel details, including type of tunnel. The protocol supports traceProbe and traceResponse messages carried over UDP.

Another interesting proposal for MPLS-OAM can be found in [31].

3.6 Current Activities in IP/MPLS Network Management

Through this chapter we reviewed several aspects of IP network management. A primary conclusion is that much unrelated work is being done in this field. A desirable trend is to somehow organize and bring together these valuable efforts to come up with some standard way of solving some of the challenges faced by this changing field. Some of the efforts related to PBNM and MPLS management have already been mentioned, and in this context it worth making reference to the standardization efforts for configuration and performance management interfaces. In this regard we will consider the relevant IETF working groups activity and then the TeleManagement Forum (TMF) IP Network Management (IPNM) initiative.

3.6.1 The IETF Working Groups

The SNMPCONF working group promotes policy-based configuration management using SNMP as the communication protocol between managers and agents. It provides considerations regarding data integrity on sensible configuration data transactions (this concept is incorporated in the management API example provided earlier).

Also currently active working groups in relation to the considered field are those grouped together under the SubIP area, mainly the MPLS working group and the TE working group. Some of the efforts undertaken in the MPLS working group have been described before, and it is worth mentioning the permanent update of the MPLS MIBs mentioned earlier. The Framework for Internet Traffic Engineering Measurement [32] is also worth to mentioning in regard to performance management for MPLS-TE.

3.6.2 TMF IPNM

The IPNM is an effort to deploy a standard common object request broker (CORBA)–based NML-EML interface for IP networks. This is described in the document TMF 611.

The objective for the IPNM team is to create solutions for the integrated management of multivendor IP networks. Solutions will enable complete flowthrough of IP network services from BML, to SML, NML, and EML.

The IPNM project provides an open and accessible venue for service providers, equipment vendors, and software vendors to collaborate and agree upon a common industry solution, which will enable integrated management of multivendor IP networks. It is a priority of the IPNM project to ensure that all recommended solutions are implementable. To facilitate a timely delivery of the recommended solutions, CORBA is used as the enabling technology.

IPNM references include recommendations from other bodies, as well as contributions from member participants. Standard bodies and groups include ITU-T and ETSI, IETF, DMTF DEN. Other source documents are listed in [33].

3.7 Summary

In the conventional best-effort Internet, there is no network model due to the characteristic connectionless nature of such a network. Therefore we could say that the IP layer is element-level-management sensitive only. Nevertheless the deployment of MPLS in the IP network has changed that status. IP MPLS-enabled networks can be characterized by a network model constituted by the end-to-end LSPs. These paths can be established under the control of a routing algorithm in the same manner as virtual paths in ATM networks. MPLS allows network administrators to specify the physical path that LSPs take across the network. Moreover these LSPs can be enforced to meet specific performance requirements and can be monitored during their life span. Therefore, this converts the IP MPLS-enabled network in a network-level manageable entity like any connection-oriented network.

Conventional or up-to-date IP network management has been based on the facilities provided by the SNMP in the areas of performance (monitoring) and fault management, whereas configuration management was implemented by means of vendor-specific mechanisms. Management based on SNMP has been supported by several MIBs, one of whose more representative cases is the MIB-II. This MIB allows for performance management of network device interfaces as well as a very limited functionality in configuration and fault management. On the other hand, IP MPLS-enabled networks can also be managed through SNMP thanks to the support of several ad hoc defined MIBs. Among these we mention the MPLS-LDP-MIB, which is intended to manage the LDP, a protocol used for distributing the labels through the network. Also worthy of mention is the MPLS-LSR-MIB and the MPLS-TE-MIB, both with functionality to carry out configuration and performance management of the LSPs. As an example of the usage of the MPLS-TE-MIB, we provide the implementation of an API to configure LSPs, making use of all the functionality provided by the device vendor (Cisco in this case).

PBNM has emerged as a new paradigm intended to abstract the service characteristics from the implementation details, thus allowing for the translation and mapping of application-level management statements to device-level management commands. Although not restricted to the IP world, PBNM has been growing in part thanks to the effort of the IETF. The IETF-RAP working group has developed a well-known framework specifying the functional elements of a

PBNM system (policy-decision points, policy-enforcement points), protocols, and schema.

We have presented a selection of management tools grouped into several categories. There are many of such tools that can be characterized as open source, and others are APIs that make use of SNMP. Into another category, we group tools supporting the PBNM paradigm. Among these tools is worthy to mention Ponder, a language and toolkit for the specification of several types of management policies. Finally, we present tools that allow for the monitoring of MPLS LSPs in IP MPLS-enabled networks.

Efforts in the all these IP management areas are currently being carried out by several IETF working groups such as the SNMPCONF working group, the MPLS working group, the TE working group, and the Telemanagement Forum (TMF). Among the initiatives of the TMF, we mention the IPNM initiative intended to define a CORBA-based network-management-level to element-management-level interface and some catalyst projects.

References

[1]   Comer, D. E., *Internetworking with TCP/IP Vol.1: Principles, Protocols, and Architecture,* Englewood Cliffs, NJ: Prentice Hall, 2000.

[2]   Rosen, E., A. Viswanathan, and R. Callon, "Multiprotocol Label Switching Architecture," IETF RFC 3031, January 2001.

[3]   Awduche, D., et al., "RSVP-TE: Extensions to RSVP for LSP Tunnels," IETF RFC 3209, December 2001.

[4]   Jamoussi, B. (ed.), "Constraint-Based LSP Setup using LDP," IETF RFC 3212, January 2002.

[5]   Internet Engineering Task Force, "Internet Traffic Engineering (tewg) Working Group," www.ietf.org/html.charters/tewg-charter.html.

[6]   Faucheur, F. (ed.), "Protocol Extensions for Support of Diff-Serv-Aware MPLS Traffic Engineering," draft-lefaucheur-diff-te-proto-02.txt, work in progress, expires April 2003.

[7]   Stallings, W., *SNMP, SNMPv2, SNMPv3, and RMON 1 and 2,* Reading, MA: Addison-Wesley, December 1998.

[8]   Nadeau, T., et al., "Multiprotocol Label Switching (MPLS) Management Overview," draft-ietf-mpls-mgmt-overview-02.txt, work in progress, expires December 2002.

[9]   Baliosian, J., and E. Grampín, "A Generic Object Oriented MPLS-TE Application Programming Interface," *UPC Internal Report,* June 2002.

[10] Cisco Systems, "CISCO-CONFIG-COPY-MIB," ftp://ftp.cisco.com/pub/mibs/v1/CISCO-CONFIG-COPY-MIB-V1SMI.my.

[11]   Cisco Systems, "How to Copy Configurations To and From Cisco Devices Using SNMP," www.cisco.com/warp/public/477/SNMP/copy_configs_snmp.shtml.

[12]   Cisco Systems, "CISCO-RTTMON-MIB," ftp://ftp.cisco.com/pub/mibs/v2/CISCO-RTTMON-MIB.

[13]   Cisco Systems, "Network Monitoring Using Cisco Service Assurance Agent," www.cisco.com/univercd/cc/td/doc/product/software/ios122/122cgcr/ffun_c/fcfprt3/fcf017.htm.

[14]   Cisco Systems, "Cisco VPN Management," www.cisco.com/warp/public/cc/pd/nemnsw/vpnm.

[15]   Kolon, M., Juniper Networks, "MPLS VPN Provisioning," www.juniper.net/techcenter/notes/552005.html.

[16]   Orchestream Web site, www.orchestream.com.

[17]   WANDL, "IP/MPLSView," www.wandl.com/html/mplsview/MPLSview_new.cfm.

[18]   Chadha, R., and Huai-An (Paul) Lin, "Policy Information Model for MPLS Traffic Engineering," draft-chadha-policy-mpls-te-00.txt, work in progress, expired January 2001.

[19]   Internet Engineering Task Force, "Resource Allocation Protocol (rap) Working Group," www.ietf.org/html.charters/rap-charter.html.

[20]   Durham, D., et al., "The COPS: Common Open Policy Service Protocol," IETF RFC 2748, January 2000.

[21]   Moore, B., et al., "Policy Core Information Model—Version 1 Specification," IETF RFC 3060, February 2001.

[22]   Distributed Management Task Force, "Common Information Model (CIM) Specification, Version 2.2," June 1999.

[23]   Oetiker, T., "Multi Router Traffic Grapher," people.ee.ethz.ch/~oetiker/webtools/mrtg/mrtg.html.

[24]   SimpleWeb, "SNMP/Management Software," www.simpleweb.org/software.

[25]   Levi, D., and J. Schoenwaelder, "Definitions of Managed Objects for the Delegation of Management Scripts," IETF RFC 2592, May 1999.

[26]   Distributed Software Engineering Department of Computing at Imperial College, London, UK, "Ponder: A Policy Language for Distributed Systems Management," www-dse.doc.ic.ac.uk/Research/policies/ponder.shtml.

[27]   Distributed Software Engineering Department of Computing at Imperial College, London, UK, "Policy Research Group," www-dse.doc.ic.ac.uk/policies.

[28]   Sloman, M., "Policy Driven Management for Distributed Systems," *Journal of Network Systems and Systems Management*, Vol. 3, December 1994, pp. 333–360.

[29]   Kompella, K. et al., "Detecting MPLS Data Plane Liveliness," draft-ietf-mpls-lsp-ping-01.txt, work in progress, expires April 2003.

[30]   Bonica, B., K. Kompella, and D. Meyer, "Tracing Requirements for Generic Tunnels," draft-ietf-ccamp-tracereq-00.txt, work in progress, expires February 2003.

[31]   Allan, D. (ed.), "A Framework for MPLS User Plane OAM," draft-allan-mpls-oam-frmwk-03.txt, work in progress, expires April 2003.

[32]   Wai Sum Lai, et al., "A Framework for Internet Traffic Engineering Measurement," draft-ietf-tewg-measure-03.txt, work in progress, expires March 2003.

[33]   IPNM references:

TMF MTNM (Multitechnology)—TMF 513 (BA), 608 (IA);

TMF 820 IP Service Management IIS V1.5;

TMF 821 IP VPN IIS V1.5;

TMF GB-909, Generic Requirements for Telecom Management Building Blocks;

TMF 801, 802, 803.

4

Management of the WDM Network Layer

4.1 Introduction

As the WDM network layer is often used as a server layer network for other technologies like IP, a coordinated if not an integrated network management is imperative to make the best use of the available resources at the establishment of end-to-end connectivity. It is also crucial to take consequent actions in WDM to overcome network faults or performance problems in client networks.

WDM transport networks have an architecture inherited from their predecessors: the SDH networks. Therefore, the management of both types of technologies follows similar trends. The network elements are organized in the lowest level of the management hierarchy and depend on network-element management systems. These management systems interact with the network devices to define their behavior, so, in order to understand the capability of such systems, it is necessary to know the functionality and characteristics associated with the different elements in the optical network. Also, on top of the network-element management systems, we find the network management systems in a TMN-oriented architecture. These systems are intended to provide end-to-end connectivity as in electrical connection-oriented networks. What is important here is to know the modeling aspects of the optical technology to understand the requirements for the corresponding network managers.

From a functional point of view, the management is also structured around the five functional management areas, namely FCAPS. Several operation systems may be working together at these functional levels based on the manager-agent paradigm. The management interface can follow well-known protocols like SNMP with the network elements or operations systems modification of intelligent network elements (OSMINE) [1] or the multitechnology network management (MTNM) [2] between the NML and the EML.

This chapter is intended to present the WDM resources to be managed, both at network-device level and at network level, as well as the technologies used for the most common management procedures. To fulfill this objective, we have the following structure: Section 4.2 summarizes the main network devices encountered in WDM networks already mentioned in Chapter 2 when describing the architecture of the OTN, while Section 4.3 presents the layered model of the OTN. Devices and entities exposed in these two sections will be the subject of management. Section 4.4 enters in detail with the management issues following a functional organization in three management areas: configuration, fault, and performance management. This sections ends with a reference to the main standards or de facto standards used in management interfaces.

4.2   The WDM Network Elements

The two most important network elements within WDM networks are the OADM and the OXC. Similarly to the SDH ADM, the OADM allows the grooming or splitting of the optical signal along the transmission path, permitting the construction of more complicated and flexible networks topologies. Consequently, a WDM ring can be constructed in a way similar to SDH rings.

The OXC network element allows the (de)multiplexing of two or more optical signals into the transmission fiber or vice versa. Optical multiplexers/demultiplexers, optical switches, and wavelength converters usually compose the OXCs. Figure 4.1 depicts the architecture of an OXC element. The incoming optical signal is demultiplexed into several wavelengths that are sent to the optical switches. The optical switches route the wavelengths to optical multiplexers, creating one optical signal suitable for transmission. Wavelength converters can be used between the optical switches and the multiplexers, providing more routing flexibility.

4.3   Optical Transport Network Modeling

Introduced in Chapter 2, OTNs provide transport services via lightpaths and can be considered the next step in the evolution from WDM point-to-point networks. A lightpath is simply a high-bandwidth pipe carrying data up to several gigabits per second. An OTN is composed of several OXCs, which are connected in a certain topology of choice. Lightpaths are set up and torn down either by the management system of the network or by means of signaling. The

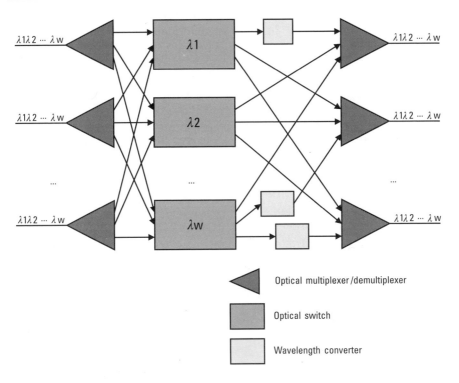

Figure 4.1 Architecture of an OXC node.

network should be also capable of monitoring the network elements and facilitating recovery from possible faults.

Some interesting features of OTNs are [3]:

- *Transparency.* Lightpaths can transfer any variety of bit rates and protocols and can be considered as protocol insensitive. They allow the support of a variety of higher layers concurrently. Some wavelength can carry SDH traffic, whereas other can carry ATM cells.

- *Wavelength reuse.* The number of the lightpaths that the network supports is greater than the number of the wavelengths, as the wavelengths can be spatially reused.

- *Reliability.* The network should be capable in the event of a fault to reroute the affected traffic over alternative paths. A high degree of reliability is of vital importance for network operators.

- *Virtual topology.* Virtual topology is the topology that is seen by the higher layers. It is a graph consisting of network nodes, with an edge between the nodes if there is a lightpath between them.

- *Circuit switching.* The lightpaths can be established and torn down in a way similar to the circuit-switched networks.

4.3.1 Functional Layers

OTNs perform different tasks to provide the transport of client-layer signals. Digital processing is required as these client layer signals are processed in the optical domain and because optical technology has its limitations on supervision and performance assessment capabilities. Prior to designing a new OTN, it is desirable to define the functional requirements of the network architecture and to assess the requirements imposed on the management system.

ITU-T recommendation G872 [4] outlines a three-layer functional architecture of an OTN, described from the network level point of view and in accordance with the terminology and conventions as stated in ITU-T G805 [5]. The OTN is, as illustrated in Figure 4.2, decomposed into independent transport layers: the optical channel (OCh), the optical multiplex section (OMS), and the optical transmission section (OTS).

4.3.1.1 OCh Layer

This layer provides end-to-end networking of optical channels (lightpaths) for transparently conveying client information of varying format (e.g., SDH STM-N, PDH 565 Mbps, and cell based ATM). To provide end-to-end networking, the following capabilities are included in the OCh layer network:

- Optical channel connection rearrangement for flexible network routing. The OCh network topology can be reconfigured by manually or automatically routing wavelengths.

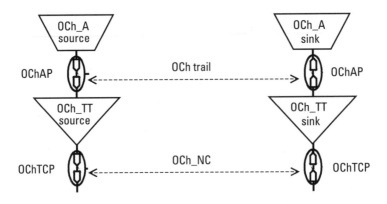

Figure 4.2 Functional architecture of the OTN (partial view).

- Optical channel overhead processes for ensuring integrity of the optical channel adapted information. For instance, the connectivity integrity is validated, the transmission quality is assessed, and transmission defects are detected and indicated. Failures detected by the OCh layer will lead to a server signal fail towards the client layer.

- Optical channel supervisory functions for enabling network-level operations and management functions, such as connection provisioning, quality of service parameter exchange, and network survivability. All supervision functions apply to the OCh layer. In case there is no possibility for rearranging network connections on the OCh layer, no supervision function for connectivity is required.

4.3.1.2 OMS Layer

This layer provides functionality for networking of a multiwavelength optical signal; this is a grouping of OChs. The capabilities of this layer include:

- Optical multiplex section overhead processes for ensuring the integrity of the multiwavelength optical multiplex section–adapted information; failures detected at the OMS layer will lead in turn to a server signal fail at the OCh layer.

- Optical multiplex section supervisory functions for enabling section-level operations and management functions, such as multiplex section survivability. In contrast to the OCh layer, the OMS layer does not have connectivity supervision function because there is a one-to-one relationship between the OTS and the OMS layers. Other supervision functions are for further study (e.g., signal quality supervision).

These networking capabilities performed for multiwavelength optical signals provide support for operation and management of optical networks.

4.3.1.3 OTS Layer

This layer provides functionality for transmission of optical signals on optical media of various types (e.g., ITU G.652, G.653, and G.655 fiber). The capabilities of this include:

- Optical transmission section overhead processing for ensuring the integrity of the optical transmission section–adapted information; characteristic information is inserted to retrieve information about the optical signal quality. This characteristic information is referred to as an optical transport module (OTM).

- Optical transmission section supervisory functions for enabling section-level operations and management functions, such as transmission section survivability.

As the OTS layer is the lowest layer of the OTN, all of its network connections are supported directly by the optical media layer. Because this optical layer does not contain active components, server failure errors will not be received, unlike the other OTN layers.

4.3.2   Optical Transport Services

Because the traffic-carrying wavelengths in a WDM-based OTN provide great bit-rate and protocol independence, such an OTN can be seen as a server layer that can carry different client signals (IP, ATM, and SDH). From this point of view, the OTN provides transport services to the upper layers. Three categories of transport services can be identified:

1. *Leased OCh service.* An OCh can be set up either using the network management resulting in a permanent set up or using signaling and routing protocols, which is more flexible.

2. *Leased wavelength.* The customer of this service should be equipped with colored line terminals.

3. *Leased dark fiber.* In this case, the provider of this service has no control of the fiber, and the customer can use it according to his or her needs.

From a carrier's point of view, the provision of lightpaths to the customers can be classified into the following categories:

- *Lightpaths that must be protected in the optical layer.* In this category, the client layer relies on the optical layer for protection and has no restoration/protection mechanism itself.

- *Lightpaths that must not be protected in the optical layer.* An example in this category is SONET/SDH signals that rely on the protection/restoration mechanism of the SONET/SDH layer.

- *Lightpaths that are indifferent to protection.* IP traffic fits in this category, because IP has its own mechanism for protection by rerouting the traffic around the affected segment of the network. However, protection at the optical layer is welcome because it takes less time to protect in the optical layer than it does rerouting in the IP.

- *Lightpaths that may be protected on a best-effort basis.* Here, protection is provided only when bandwidth is available in the optical layer.

In the event of a fault, affected lightpaths will use lower priority lightpaths to resume their transmission and, hence, preempt the traffic on these lower priority lightpaths.

4.4 Managing WDM Networks

Initially, the WDM network merely provided point-to-point connectivity with static routes and network topology. For this type of WDM network configuration, the management is limited to network planning issues. In order to establish end-to-end connectivity, fiber connections need to be changed. Now, more advanced configurable WDM networks can be created by using OXC network elements. As cross connections can be made between a port on the OXC and the multiwavelength fiber, wavelengths could either be switched from one end to the other of the network, or wavelength translation can be applied to have the information transported through the network by different wavelengths. This configuration allows for the routing of WDM wavelengths through the network and imposes requirements on the network management system for appropriately setting the OXCs [6, 7]. WDM networks are hierarchically organized. For example, a WDM NMS manages several element management systems (EMSs), which again each manages a group of WDM network elements. As the functionality for protection and restoration is gradually incorporated due to the greater extent of reconfigurability, the need for the SONET/SDH intermediate layer is eliminated. Additionally, parts of the OAM functions of SDH have been incorporated in WDM networks. The next generation of WDM infrastructure uses packet switching instead of wavelengths. This means that data packets, like IP, can be directly transported across the network. Within the network, these packets are appropriately switched to reach their destination. Consequently, a much finer granularity in traffic multiplexing is reached. Obviously, more intelligence will be necessary in the network elements, which will simplify the network operations. The following sections will present the various issues involved in configuration, fault, and performance management in configurable WDM networks.

4.4.1 Configuration Management

In general, configuration management in WDM performs two different tasks; connection management for facilitating the wavelength connection set up and tear down and network inventory management for maintaining a logical and physical representation of the network. Before wavelengths are set up, the management system has to determine an appropriate route through the network that complies with QoS and routing restrictions. Subsequently, the route is implemented instructing the WDM network elements or EMSs to create the cross connections.

One of the most important tasks that can be performed by the management system of the network is to define the routes that the traffic will follow. In WDM networks, where connections are supported by lightpaths, there are two categories of wavelength routing:

1. *Static routing.* This applies to a network configuration where no wavelength conversion is possible, which means that one wavelength is required to connect both ends. Thus, the routing is fairly simple, and the path through the network is predetermined.

2. *Dynamic routing.* This type of network is characterized by its ability to perform wavelength conversions and thus does not require a single wavelength to connect end points. Routing in this type of network is also referred to as reconfigurable routing, and requires reconfigurability of the WDM equipment such that a path can be computed to fulfill the desired characteristics. Applying wavelength conversion, the connection blocking probability is reduced.

Given the network topology and a set of end-to-end connection requests, the routing and wavelength assignment (RWA) problem, studied by [8] and others, defines the problem of determining a route and the wavelength assignment for each route request, while at the same time minimizing the number of the wavelengths. An important factor, besides the network capabilities, in the RWA problem is what type of connection requests is issued. In case of *static traffic*, the connection requests are known in advance, while in case of *dynamic traffic*, connection requests arrive randomly. The RWA problem for static traffic is also referred to as the static lightpath establishment (SLE) problem. In general, the RWA problem is tightly coupled with the virtual (logical) topology design. The latter resolves the connectivity between different nodes, taking into consideration a set of restrictions like traffic demands matrices, nodal resources, policy constraints, and overprovision constraints. However, it can be assumed that the virtual topology design generally deals with long-term concerns, and it provides input to the RWA problem. The RWA problem itself can be further decomposed into two subproblems: route selection and wavelength assignment.

The RWA problem is further complicated by different constraints, such as:

- *Wavelength continuity.* Two connections with the same wavelength cannot be supported on a given fiber. This constraint is of importance in networks that do not have any wavelength conversion capabilities.

- *Limited wavelength conversion capabilities.* Not all wavelengths can be converted to any other wavelength at any OXC; frequently the conversion capabilities of OXCs are limited.

- *Physical impairments.* These constraints are associated with the degradations that the optical signals could suffer due to either linear or nonlinear effects.

Many optimization schemes have been proposed [9, 10], which try to minimize various cost functions such as the number of wavelength, blocking probabilities for different network capabilities, and traffic types. Usually, a complete set of desired source-destination requests is given and routing decisions are made for all source-destination pairs in order to achieve a global optimum solution. This solution requires complicated computation, and it becomes excessively time consuming as the network size increases. Therefore, it can be used only for static traffic requests.

To cope with real-time requests, dynamic traffic that may arrive at random times, dynamic routing schemes have been proposed. These approaches lead to greater blocking probabilities, because channels are routed as they arrive. Channel rerouting is then needed in order to optimize the network utilization. This is done by reallocating the network resources. Currently, there is no standardized RWA algorithm.

4.4.2 Fault Management

Network survivability is an important factor in the design of a network, and network operators take special actions in order to minimize the disruption of the network traffic in case of a fiber defect or an equipment failure. Taking into consideration that a fiber cut affects thousand of connections, the existence of a flexible, efficient, and robust protection/restoration mechanism is of primary importance.

Protection switching in the optical layer has many similarities with the SONET/SDH approach and achieves fast restoration/protection within 50 msec. Protection can be performed either at an optical channel level, protecting individual lightpaths, or at the aggregate signal level, which corresponds to the OMS.

We can further distinguish protection at the link or the connection/path level. At the link level, the traffic is diverted where the affected link is located, whereas at the connection level, the protection occurs between the two end points of the path. The following sections describe these in more detail.

4.4.2.1   Link Protection/Restoration

In this case the protection or back-up path is disjointed from the working path at the particular link over which protection is required. When the protected link fails, traffic on the working path is switched over to the protection path. This is a local repair method that can be potentially fast. Figure 4.3 illustrates this principle; a working fiber is used for transferring the information, while a counter-propagating protection fiber is dedicated to protection purposes. Signals added to the ring are split across both fibers, and the destination node will take the best signal of both.

At the OCh layer, similarly to the SDH protection, we have the following schemes (for ring topologies) [11]:

- The optical channel dedicated protection ring (OChDPRING) is a 1+1 protection mechanism that uses two physically separated media (e.g., different fibers that for instance transport traffic in the opposite direction of the ring). In case of a two-fiber ring network, one fiber will carry the working signal while the counteracting fiber will carry the protection signal at the same wavelength. This means that the traffic entering the ring will be copied and transported twice across the ring. The destination node of the ring will select the best signal.

- The optical channel shared protection ring (OChSPRING) can be considered a 1:1 subnetwork connection (SNC) protection mechanism. In a two-fiber ring architecture, one fiber will carry on half of the available OCh space working OCh signals, and the other half is used to protect

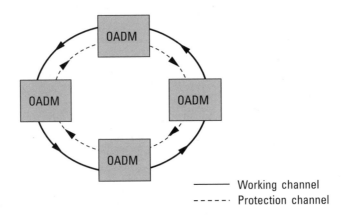

Figure 4.3 Link protection in an optical ring.

the working channels of the counteracting fiber. The counteracting fiber will have half of its OCh space allocated the other way around. Contrary to OChDPRing protection, this mechanism does not transport the working and the protection channels in normal (nonfailure) circumstances.

As explained earlier, the OCh is just one of the many channels that might be using the same fiber, whereas OMS protection aims at protecting a bundle of optical channels.

At the OMS layer we have the following schemes (again for ring topologies):

- Optical multiplexing section dedicated protection ring (OMSDPRING), where one fiber on the ring is used for working traffic and the other fiber is used for protection traffic;
- Optical multiplexing section sharing protection ring (OMSSPRING), where, in a two-fiber OMSSPRING, half of the wavelengths on each fiber are reserved as protection channels. In the event of a failure condition, the nodes adjacent to the failure will loop back all of the affected lightpaths at once on the protection channels of the ring.

4.4.2.2 Connection Protection/Restoration

In the case of network connection protection, the end points of the network connection are protected. Under network connection protection, the protection path is completely separate from the working path. The advantage of network connection protection is that the protection path protects the working path from all possible link and node failures along the path, except for failures that might occur at the ingress and egress nodes. The optical subnetwork connection protection (O-SNCP) belongs to this category. This type of protection is implemented by sending the signal across two different routes through the network. It concerns not individual links but a sequence of links and nodes.

4.4.2.3 Linear Protection/Restoration Schemes

Linear protection can take place through the following alternatives at the OMS level:

- *1+1 protection.* For this type of protection scheme, the optical signals are always split at the transmitting node and traverse the network via different paths. Half of the fibers within the network are dedicated for protection purposes. The receiving node "listens" to both and will select the best.

- *1:1 protection.* This protection scheme only uses the protection path as soon as the working path fails. Then the protection path will take over from the working one. Usually, the protection path is used for low-priority traffic that will be dropped in case the working path fails. Because the transmitting node must be notified of the working path failure, this scheme is considerably slower than the first scheme.

- *1:N protection.* This scheme is different from the 1:1 protection with respect to the number of protection paths per working path. In this case, each protection path is used to protect *N* working paths. If a failure occurs with a certain working path, the protection path will take over until the working path has recovered. Similar to 1:1 protection, the protection path can be used for carrying low-priority traffic. Note that if there is a second failure, there will be no protection path to take over.

- *M:N protection.* For each of the *N* working paths in this scheme, there are *M* protection paths available. The remaining characteristics are similar to 1:*N* protection.

4.4.3   Performance Management

The need for monitoring and managing the performance of networks is especially acute in WDM networks because of the high bandwidths per wavelength and the number of users affected in case of problems. Performance monitoring in WDM typically involves the monitoring of the optical signal quality, determined by the power level and the signal-to-noise ratio. However, there are difficulties relating the signal quality with the performance management metrics as bit error rate (BER) and errored seconds (ES) [12]. In addition to monitoring the optical signal nonintrusively, three more requirements are identified [13] that should be met by a performance management method for all-optical WDM networks:

- It should not need to access a specific data frame (frame format transparency).

- It should work at a wide variety of bit rates (bit rate transparency).

- Identification of the strength and source (type) of signal degradation should be possible.

In the following section, performance-management methods are identified that are relevant for all-optical WDM networks.

4.4.3.1 Spectral Analysis

Spectral analysis is the main method used for performance management, because it only monitors the parameters of the optical carrier alone, without concern for the modulated signal itself. Hence, it is immediately compatible with all of the requirements for performance management in all-optical networks. The method consists of decoupling a sample of the optical carrier from the transmission fiber and feeding it to an optical spectral analyzer. Here, the results of the analysis are compared against a predefined set of results defined for correct operation. The supervision is carried out at the various network elements. For example, at a switching node every input and output fiber is monitored by the network analyzer. The analyzer uses a selector mechanism to get a sample of the optical signal from every port one at a time. This method only allows the measurement of power-related figures such as power per carrier, carrier wavelength, and optical signal-to-noise ratio. While the first two characteristics can be measured quite accurately using this method, the optical signal-to-noise ratio may only be an estimate of the real value due to several different factors, such as the presence of four wave mixing or cross phase modulation or even the presence of demultiplexers or other optical filters [14].

The performance of the network depends directly on a number of the optical carrier parameters. If one (or more) of these parameters is found to be below (or different) from given thresholds, the performance of the network may be severely affected (depending on the extent of the degradation), hampering its service quality as seen by the users. The main optical carrier parameters that must be monitored in a WDM network are:

- The number of carriers, indicating an unbroken path between the carrier generation and measurement point—this could indicate that more than one carrier is used at the same frequency;
- Total optical power, defined as the sum of all carrier power plus noise;
- Carrier identification, which helps identify the carrier by manipulating some of the carrier parameters at certain network elements;
- Optical power per carrier, measurements indicate whether the optical power is within the predefined range;
- Carrier-to-noise ratio, per carrier, indicating that the noise accumulated along the transmission path does not exceed the value beyond which the transmission is no longer error free;
- Center wavelength drift, defined as the difference between the nominal carrier wavelength and its actual value;

- Spurious emission, which usually indicates a laser malfunction and may cause interference with other carriers;
- Subsidiary optical carrier parameters (other optical parameters that may affect upon one or several main optical carrier parameters).

4.4.3.2   Signal Quality Supervision Methods

The optical spectral analysis technique also has important limitations in its application to the performance management of all-optical transport networks. Its limitations stem from the fact that this method only monitors the optical carrier. Therefore, alternative supervision methods may be required in order to also monitor the optical signal quality itself (i.e., the quality of the modulated signal), while still conforming to the requirements for performance management.

The optical modulated signal quality can be affected by several different optical parameters. Some of these parameters are essentially static in nature (i.e., their values change very slowly during the network's life span). In these cases, the network is engineered so as to guarantee that these parameters fall within acceptable limits. In other cases, hardware countermeasures might be in place in the network to guarantee that the value of these parameters falls within acceptable limits [15, 16].

Nevertheless, in agreement with the third requirement given earlier, supervision of these parameters may be required as their value slowly drifts away (e.g., due to component aging).

The relevant parameters affecting the signal quality are:

- Dispersion, defined as the spreading of an optical pulse as it propagates along its path;
- Crosstalk, which is an interference with the desired signal caused by a spurious signal falling at the same frequency of the signal;
- Bit error counting, which consists of applying an error detection pattern to the payload in the transmitter side, which is sent as overhead to the receiver—this error detection pattern will allow the error counting at the receiver side;
- Amplitude distribution of the optical signal, which produces a histogram of the distribution of the actual signal amplitudes corresponding to both the logical "one" and "zero" symbols. From the recorded histogram, it is possible to calculate an estimate for the BER of the transmission in a given path [17].

As with fault management in WDM, performance management is further complicated when multilayer networks are involved (e.g., SONET), thus requiring integration of the management functions.

4.4.4 Management Interfaces

WDM network elements are not tied to any particular management interface; this is also true for WDM element and network management systems. Possible interfaces may range from proprietary to standardized interfaces. This section will describe a set of frequently used standardized interfaces for managing WDM network elements. The main management interfaces competing on the network element layer are TL1, SNMP, and common management information protocol (CMIP). Also Telcordia's OSMINE is used in networks as a means to communicate between the OSS and various types of network elements.

Transaction language 1 (TL1) [18] is a text-based, command-line, telecommunications protocol for network element management, specified by Bellcore in 1986. A large number of technologies and vendors have network elements that are managed via TL1. General acceptance of TL1 was brought with the reliance of operators on Telcordia operation support system (OSSs) for interfacing with network elements. In today's large OTNs, TL1 is the dominating protocol, as SNMP is for packet networks.

SNMP is the IETF's standardized network management interface and dominates the packet networks. While the majority of the WDM equipment vendors mainly offer a TL1 interface, some do also offer SNMP. However, limited security [19] is built into SNMP, and it is vulnerable to malicious individuals due to its inconsistent behavior when unexpected input is provided. The OSI network management model proposes a different approach than the one described earlier. Manager and agent communicate by means of the CMIP protocol [20, 21]. This is an alternative that offers richer functionality than TL1 but is more costly to deploy. Unlike TL1, CMIP and SNMP are both based on platform-centric manager-agent paradigms, each with a different information model. This is in contrast to TL1, which does not have any information model concept defined.

Different from the previous models, the OSMINE [1] is a process that, according to Telcordia, integrates equipment manufacturer's products with Telcordia OSSs. OSMINE, that is conforming the TL1 interface, gives network equipment interoperability with Telcordia OSS systems. Telcordia offers this process as a service to optical equipment vendors to certify the equipment. Because Telcordia is in a monopolistic position with the regional Bell operating companies (RBOCS), other vendors' equipment must be made OSMINE compliant. OSMINE is an ongoing process because Telcordia is the single authority to determine the scope of work for OSMINE certification. Network element vendors that intend to deploy products at multiple service providers may be required to undergo OSMINE certification multiple times. Sometimes the EMS or NMS that come with a vendor's network element must undergo OSMINE certification also. Moreover, when a new Telcordia OSS system is introduced into an existing SP network, certification of the network elements may be necessary.

TMF is working in this direction by launching programs and catalyst projects to capture the needs of network operators and service providers willing to enable the "technology integration" and the end-to-end "process automation" of telecommunications and data services operations. In this context, TMF has adopted and started specifying transport-technology-independent common management interfaces from the EML towards the NML and from the NML towards the SML, sometimes called the *open CORBA* interfaces. The outcome of these efforts was a series of documents providing, among others, the MTNM interface [2] and the connection and service information model (CaSMIM) specifications. The documents provided have a specific methodology, first stating the problem, analyzing the requirements and the use cases of the proposed solution (business agreement), then specifying the protocol neutral (information agreement) and protocol specific (solution set) information models, and finally implementing the solution sets (interface implementation specification) with the catalyst projects. There are several transport technologies under consideration with the focus of attention given to WDM, as it will soon be dominant in transport networks.

4.5 Summary

The WDM network is constituted by network elements like optical multiplexers, optical switches, and wavelength converters, which in turn are the constituting devices of the optical cross connects. The setup, monitoring, and maintenance of these devices will be the main scope of network-element management systems.

On the other hand, the WDM network follows a model that is divided into three layers, the OCh, the OMS, and the OTS. Each of these layers will provide functionality to allow operation and management, and therefore these three entities will be under the scope of the NMSs. On the other hand, configuration, fault, and performance management are the functionality covered in this type of network.

Configuration management facilitates the set up and tear down of optical channels as well as the network inventory management. Optical channels are established according to routing algorithms that can be static or dynamic depending whether the same wavelength must be used between the optical channel end points. Currently, there is no standardized RWA algorithm for determining the route and the wavelength for the network connectivity requests. Owing to the high capacity of links in WDM networks, the existence of a flexible and efficient protection/restoration mechanism is the primary aim in the area of fault management. In the case of link protection, the backup path is disjointed from the working path at the particular link where protection is required. Also, with regard to network connection protection, the protection path is disjointed from the protected one. Performance management is based on the monitoring of relevant optical carriers characteristics, such as the total number of carriers, the total optical power, and for each carrier the power-to-noise ratio, the center wavelength drift, spurious emissions, and the optical signal quality itself (e.g., dispersion or crosstalk).

As far as management interfaces is concerned, TL1, SNMP, and CMIP are the most common ones at the network element level, whereas Telcordia's OSMINE and the TMF MTNM are the most representative at the interface between network management systems and network-element management systems.

References

[1] Communications Industry Researchers (CIR), "RBOC, Optics and OSMINE: The Latest Thinking," Vol. 1, No. 21, 2001.

[2] TeleManagement Forum, TMF513, "Multi-Technology Network Management Business Agreement," Vol. 2.1, September 2002.

[3] Communications Industry Researchers (CIR), "Optical Networking Watch," Vol. 2, No. 21, Issue 44, 2002.

[4] ITU-T Recommendation G.872, "Architecture of Optical Transport Networks," November 2001.

[5] ITU-T Recommendation G.805, "Generic Functional Architecture of Transport Networks," March 2000.

[6] Wei, J. Y., et al., "Network Control and Management of Reconfigurable WDM All-Optical Network," *Network Operations and Management Symposium*, NOMS 98, IEEE, Vol. 3, 1998, pp. 880–889.

[7] Wei, J. Y., "Advances in the Management and Control of Optical Internet," *IEEE Journal on Selected Areas in Communications*, Vol. 20, No. 4, May 2002, pp. 768–785.

[8] Zang, H., J. P. Jue, and B. Mukherjee, "A Review of Routing and Wavelength Assignment Approaches for Wavelength-Routed Optical WDM Networks," *Optical Networks Magazine*, Vol. 1, No. 1, January 2000, pp. 47–60.

[9] Lee, T., and K. Lee, "Optimal Routing and Wavelength Assignment in WDM Ring Networks," *IEEE Journal on Selected Areas in Communications*, Vol. 18, No. 10, 2000, pp. 2146–2154.

[10] Huang H., and J. A. Copeland, "Hybrid Wavelength and Sub-Wavelength Routed Optical Networks," *Proc. IEEE Globecom*, Vol. 4, 2001, pp. 2119–2123.

[11] Telcordia document, "BR-GR-2979-CORE 04 Generic Requirements for Optical Add-Drop Multiplexers (OADMs) and Optical Terminal Multiplexers (OTMs)," December 2001.

[12] Wilson, B. J., et al., "Multiwavelength Optical Networking Management and Control," *Journal of Lightwave Technology*, Vol. 18, No. 12, 2000, pp. 2038–2057.

[13] Hannik, N., et al., "Application of Amplitude Histograms to Monitor Performance of Optical Channels," *IEE Electronics Letters*, Vol. 35, No. 5, 1999, pp. 403–404.

[14] Heppner, B. H., and M. Bischoff, "Performance Management in WDM Networks," WTC2000 Presentation, www.wtc2000.org/pdf/ab_109s.pdf, December 2002.

[15] Boroditsky, M., et al., "Power Management for Enhanced System Performance of Passive Optical Rings," *LEOS 2000*, Rio Grand, Puerto Rico, 2000, pp. 226–227.

[16] Fei, Y., et al., "A Novel Scheme of Power Equalisation and Power Management in WDM All-Optical Networks," *IEEE Photonics Technology Letters*, Vol. 11, No. 9, 1999, pp. 1189–1191.

[17] Downie, J. D., and D. J. Tebben, "Performance Monitoring of Optical Networks with Synchronous and Asynchronous Sampling," *Proc. Optical Fiber Communications Conference and Exhibit*, Vol. 3, Anaheim, CA, March 17–22, 2001, WDD50–WD1-3.

[18] TL1—the Telecom Management Protocol, www.tl1.com, December 2002.

[19] Davis, A., et al., "Understanding the Risks of SNMP Vulnerabilities," www.lucent.com/livelink/255868_Whitepaper.pdf, December 2002.

[20] ITU-T Recommendation X.710, "Information Technology—Open Systems Interconnection—Common Management Information Service," October 1997.

[21] ITU-T Recommendation X.711, "Information Technology—Open Systems Interconnection—Common Management Information Protocol: Specification," October 1997.

5

IP Over WDM Integration Mechanisms

5.1 Introduction

Different approaches have been proposed for the smooth, fast, and reliable provisioning and management of Internet services over the optical layer. The approaches can be categorized in three main areas: ones using the control plane only, ones using the management plane only, and ones combing the management and control plane approaches. Most of the research efforts are trying to benefit from the control and signaling mechanisms of the control plane approach in the optical layer, leaving the management functions in a supportive/secondary role. The basic idea adopted was to extend the control and signaling mechanisms of the Internet to the optical layer, delegating extra intelligence to the optical network elements (ONEs). Such efforts, driven by different standardization bodies, are among others the ITU-T automatic switched optical network/automatic switched transport network (ASON/ASTN), the Optical Interworking Forum (OIF), optical user network interface (UNI), and network to network interface (NNI) activities, and the IETF generalized MPLS framework and corresponding protocol extensions. Another integration approach would be possible through the extension of the telecom-style network management approach to the IP layer as a result of MPLS capabilities, which are similar to the connection-oriented technologies. In such a case, the integration of the IP/MPLS and WDM layers is mainly performed with management means capable of performing integrated provisioning of LSPs over optical channels (lambdas), as well as integrated multilayer fault and performance management. This approach can be considered as a midterm solution until the ground for the control mechanisms is prepared, both technically and economically.

This chapter is structured around three main sections. After this introduction, Section 5.2 is devoted to highlighting the different approaches that can be adopted in the data plane to transmit IP packets to the optical WDM layer. In Section 5.3, we detail different initiatives that are being pursued in order to get an efficient control plane supporting the integration of IP over WDM. Special attention is devoted to the IETF and ITU-T initiatives. Finally, in Section 5.4 we briefly highlight the integration of the management plane approach that is extensively described in the following chapters of the book.

5.2 IP Over WDM—The Data Plane Perspective

Different encapsulation methods have been proposed for the smooth integration of IP over WDM. The basic approaches are analyzed next, starting from IP over ATM over SDH over WDM to IP over GbE over WDM, and now IP over 10GbE over WDM. New approaches like optical packet switching are also highlighted.

5.2.1 IP Over ATM Over SDH for WDM Transmission

There are many flavors of IP over ATM (e.g., classical IP over ATM, LAN emulation, and multiprotocol over ATM). For long-haul transport over WDM, the most standard transmission format currently is to use the SDH frame. Figure 5.1 shows a possible IP/WDM network architecture that uses IP over ATM over SDH encapsulation.

In this scenario, IP packets are segmented into ATM cells and assigned different virtual connections by the SDH/ATM line card in the IP router. The ATM cells are then packed into an SDH frame, which can be sent either to an ATM switch or directly to a WDM transponder (simplified in Figure 5.1 to a single OADM ring).

At present, one of the ways to try to ensure a given QoS for an IP service is to guarantee a fixed bandwidth between pairs of IP routers for each customer (layer 2 QoS management). ATM provides a way to do so with variable granularity by means of permanent virtual channels (PVC) using the ATM management system or switched virtual channels (SVC) dynamically set up. It can also use statistical multiplexing to allow certain users to access extra bandwidth for short bursts. This can help to guarantee a fixed and arbitrary bandwidth from less than 1 Mbps to several hundred megabits per second to many different customers. In addition, the fine granularity can enable IP routers to be easily connected into a logical mesh, thus minimizing delays from intermediate routers. Another benefit in using ATM protocol is the possibility of having differentiated traffic contracts, which offer various QoS depending on the application

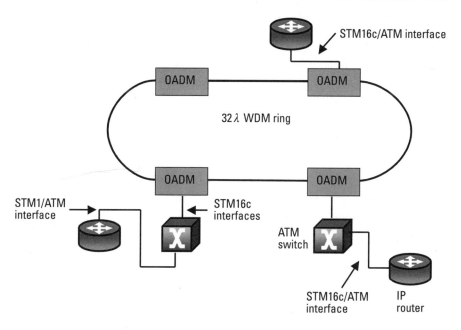

Figure 5.1 Example of IP over ATM over SDH encapsulation for transport over a WDM network.

requirements. For IP traffic, which is by essence connectionless, the unspecified bit rate (UBR) traffic contract is mainly used within ATM networks. Nevertheless, if IP applications require a particular QoS, especially for real-time constraints, it is possible to use other ATM transfer capability (ATC) such as constant bit rate (CBR) or variable bit rate–real time (VBR-rt). However, mapping variable length IP-packets onto fixed length ATM cells imposes additional overhead if fragmentation of the packet is necessary. This is sometimes called *cell tax*. Differences in size may also raise the need to pad empty space in the cells, which also gives additional overhead. One solution to prevent padding is to put packets directly one after another, but this means a potential risk of losing two consecutive packets in case of cell loss. IP over ATM can also be implemented through the use of MPLS. In this situation, the PVCs are not set up by intervention of the ATM management system, but dynamically by the MPLS protocol. For MPLS based on ATM, the label can be stored in the ATM VCI.

5.2.2 IP Over ATM Directly on WDM

It is possible to have a scenario where ATM cells are transported directly on a WDM channel. From an architectural point of view, this scenario is the same as the previous one but the ATM cells are not encapsulated into SDH frames.

Instead they are sent directly on the physical medium by using an ATM cell-based physical layer. Cell-based physical layer is a relatively new technique for ATM transport that has been developed specifically to carry the ATM protocol; this technique cannot support any other protocol except if these protocols are emulated over ATM. Some benefits of using a cell-based interface instead of SDH are:

- Simple transmission technique for ATM cells, as cells are directly sent over the physical medium after scrambling;
- Lower physical layer overhead (around 16 times lower than SDH);
- As ATM is asynchronous, there is no stringent timing mechanism to be put on the network.

However, the drawbacks are that the overhead (i.e., cell tax) is the same as for transport on SDH, the technology has not been endorsed yet by the industry, and this transmission technique can carry only ATM cells. ATM cell-based physical layers are defined by several standardization bodies. ITU-T covers the transmission rates of 155 Mbps and 622 Mbps and recently the ATM Forum completed the specification for 622 Mbps and 2,488 Mbps.

5.2.3 IP Over SDH; Packet Over SONET

It is possible to simply use SDH formats to frame encapsulated IP packets for transmission over WDM, probably using a transponder (wavelength adapter). It is also possible to transport the SDH-framed IP over an SDH transport network along with other traffic, which may then use WDM links. A scenario for transporting IP in an SDH frame over WDM is shown in Figure 5.2.

SDH can currently be used to protect IP traffic links against cable breaks by automatic protection switching (APS) in a variety of guises. The line card in the IP router performs the PPP/HDLC framing. The optical signal is then suitable for transmission over optical fiber either into an SDH network element, a neighboring IP router, or a WDM transponder for further transmission. There are also different types of IP over SDH interfaces:

- VC4 or concatenated VC4 *fat pipes*, which provide aggregate bandwidth without any partitions between different IP services that may exist within the packet stream;
- Channelized interfaces, where an STM16 optical output may contain 16 individual VC4s, with a possible service separation for each VC4. The different VC4s can then also be routed by an SDH network to different destination routers.

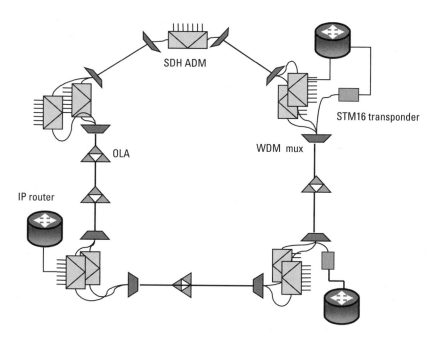

Figure 5.2 Example of IP over SDH over WDM network.

The version of IP over SDH examined here uses PPP encapsulation and HDLC framing. This is also known as POS or packet over SONET. PPP is a standardized way to encapsulate IP and other types of packets for transmission over many media from analog phone lines to SDH. It also includes functionality to set up and close links (LCP). HDLC is the International Organization for Standardization (ISO)–standardized version of SDLC, a protocol developed by IBM in the 1970s. The HDLC framing contains delimiting flag sequences at the start and end of the frame and also has a CRC checksum field for error control.

5.2.4 IP Over SDL Directly Over WDM

Simple data link (SDL) is a framing method proposed by Lucent Technologies, Inc., and can replace HDLC framing for PPP-encapsulated packets. Compared with the HDLC frame, the SDL frame has no delimiting flag sequences. Instead, the SDL frame is started with a packet-length field. This is advantageous at high bit rates where synchronization with the flag sequence is difficult. The SDL format can be inserted into an SDH payload for transmission over WDM. The SDL format can also be encoded directly onto an optical carrier: SDL specifies the bare minimum functionality to be able to do this.

SDL uses a 4-octet header, which includes the packet length, as shown in Figure 5.3. The packet can be up to 65,535 octets long. Additional error checking codes (CRC-16 or CRC-32) can optionally be used for the packet and may be placed after the packet. All bits except for the header are scrambled using an x48 scrambler. The sender and receiver scramblers are kept synchronized by occasional transmission of special packets.

SDL does not include any extra bytes dedicated to protection switching protocols (like the K1 and K2 bytes of SDH). The use of the optional payload CRCs could enable BER monitoring.

5.2.5   IP Over GbE Over WDM

The new GbE standard can be used to extend high-capacity LANs to MANs and maybe even WANs, using gigabit line cards on IP routers, which can cost five times less than SDH line cards with similar capacity. For this reason, GbE could be a very attractive means to transport IP over metropolitan WDM rings, for example, or even over longer WDM links. Furthermore, 10-Gbps Ethernet ports are likely to be standardized in the near future.

Figure 5.4 shows an example of an IP network based on GbE interfaces. The GbE line cards may be used on IP routers only, or fast layer 2 Ethernet switches may also be used to network several IP routers together.

Lower bit rate Ethernet networks (e.g., 10Base-T or 100Base-T) have been used a lot in a half-duplex mode, where the bandwidth available for transmission is shared between all users and between both directions of transmission. To policy access to the shared bandwidth, CSMA-CD is used. This imposes limits on the physical size of the network, where the transit time cannot exceed the *slot time,* which is the minimum frame length (512 bits for 10Base-T and 100Base-T). For a bit rate of 1 Gbps, using a minimum frame length of 512 bits would imply an Ethernet network only roughly 10m long. For this reason, the minimum frame length has been redefined to be 4,096 bits for GbE. However, this still limits the network size to 100m, so full-duplex mode is more attractive when using GbE.

When GbE (1000Base-X) is used in full-duplex mode, it becomes simply an encapsulation and framing method for IP packets, and the CSMA-CD functionality is not used. Ethernet switches can also be used to extend the network topology beyond a point-to-point link.

Figure 5.3 SDL header structure.

To see the full line of Artech House books and software, visit our online bookstore at:

www.artechhouse.com

Special Offers • Sample Chapters • Secure Ordering

To receive information on new and forthcoming titles from Artech House, please fill out the other side of this card and mail or fax it to one of the locations below:

For Europe, Asia, Middle East, Africa:

Artech House Books
46 Gillingham Street
London, SW1V 1AH U.K.
+44 (0)20 7596 8750
FAX: +44 (0)20 7630 0166
artech-uk@artechhouse.com

All other regions:

Artech House Publishers
685 Canton Street
Norwood, MA 02062 U.S.A.
1-781-769-9750
1-800-225-9977 (continental U.S. only)
FAX: 1-781-769-6334
artech@artechhouse.com

 Artech House Publishers   BOSTON • LONDON

Free Subscription

Artech Direct email newsletter

New Title News • Special Offers • Author Insights

[ ] Yes! Please enter my free subscription to *Artech Direct* and keep me up-to-date with emailed news of product and service information from Artech House/Horizon House Publishers.

email address: _____

You may also make my email address available to selected industry organizations and companies. [ ] Yes [ ] No

Please indicate your areas of interest

[ ] Telecommunications/Wireless/Networking

[ ] Software Engineering/Computer Security

[ ] Microwave

[ ] Radar/Remote Sensing/Electronic Defense

[ ] Signal Processing

[ ] Sensors/MEMS

[ ] Antennas & Propagation

[ ] Engineering Management

Mailing address:

Name: _____

Company: _____

Address: _____

_____

Fax or mail this card to the Artech House office nearest you. Please see other side.

A|H Artech House Publishers BOSTON • LONDON

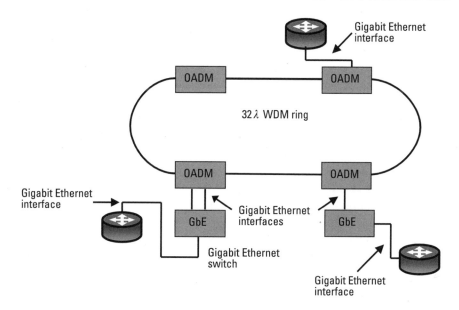

Figure 5.4 Example of IP being transported over a WDM ring with GbE framing.

GbE provides some support of CoS as defined in the standards IEEE 802.1Q and 802.1p. These standards facilitate CoS over Ethernet by providing a means for *tagging* packets with an indication of the priority or class of service desired for the packet. These tags allow applications to communicate the priority of packets to internetworking devices. RSVP or DiffServ support may also be achieved by mapping into 802.1p service classes.

5.3   Control Plane Integration

Control plane (CP) is used in the literature to refer to the set of real-time mechanisms and algorithms needed for call or connection control. It deals mainly with the signaling to set up, supervise, and release calls and connections [1]. Although a detailed decomposition of the control plane and a description of each component is not the purpose of this book, we can safely assume that the signaling protocols for connection setup, the routing protocols supporting network discovery, and the protection/recovery mechanisms are the most significant features of the control plane. In this way, it is easier to track all of the recent control plane advances and proposals about the integration of multiple layers such as IP, ATM, SDH, and WDM.

A significant element in the IP and optical integration is the corresponding business model proposed by each framework. The three basic business models described in [2], along with the requirements and issues that they impose [3], are summarized next:

- *The overlay model.* The routing algorithm, topology distribution, and connection setup signaling protocols of the IP and the WDM networks are independent. The overlay model is the one that allows an easy migration from the existing situation to the deployment of ONEs for the transport of the IP directly over WDM. However, the implementation complexity of this model is a burden, and it does not promote the integration of the control plane of the IP and the WDM networks. Only a formal request is passed from the client layer to the server layer.

- *The peer model.* The IP network has full topological view of the optical network and just a single routing algorithm instance is running in both the IP and the WDM networks. This model promotes the integration of the control plane of the IP and the WDM networks and is simpler in implementation, but its operation is far more complex than the overlay. In addition, this model can work only in cases where there is a single entity operating and managing the IP and the optical administrative domains.

- *The augmented model.* This is a combination of the previous two models. Each layer has its own protocols; however, routing information exchange is allowed between the two layers. This model can be seen as the golden mean, combining the advantages of the peer and overlay model and minimizing their disadvantages at the same time.

Some of the most significant control plane efforts on IP over optical area are reported hereafter, and although such efforts are still under development, their first results are being elaborated.

5.3.1 MPLambdaS–Generalized MPLS

The IETF has originally proposed the MPLambdaS framework [4], which extends the MPLS ideas to the optical domain, allowing the reuse of the existing Internet protocols with the appropriate extensions. The OSPF, as well as the IS-IS routing protocols, have been enhanced to disseminate information relevant to the optical domain [5, 6] allowing the route computation and automatic topology discovery. Furthermore, [7] and [8] present the mapping between the signaling messages defined in the existing IP/MPLS signaling protocols, namely RSVP-TE and CR-LDP. This enables the automation of paths setup.

MPlambdaS was originally based on the peer model, enabling direct interaction and integrated routing among the IP and the WDM layers. Later on, the IETF extended the MPLambdaS framework, which was limited to MPLS/WDM interaction, to multiple layers by means of the generalized MPLS (GMPLS). In this sense, LSRs can take forwarding or cross-connection decisions based not only on packets and wavelengths, but also on time slots (SDH layer) and physical ports (fiber layer). In [9], four classes of interfaces on LSRs are defined—the packet-switch capable (PSC), the TDM capable, the Lambda-switch capable (LSC), and finally the fiber-switch capable (FSC). Using label stacking, it is possible to make a forwarding hierarchy incorporating these classes of interfaces. Figure 5.5 presents the GMPLS hierarchy.

MPLambdaS is focused in the provisioning of MPLS electrical LSPs over optical LSPs, and it was extended through GMPLS to multiple layers, offering electrical LSPs over SDH paths over optical LSPs over cross-connected fibers. Thus, it became a superset of ASON (an ITU-T initiative), which focuses only in the optical layer as will be described later. It should be noted that although GMPLS was defined to be generic, it appears that it is being developed as an extension of particular protocols, namely RSVP-TE and CR-LDP.

5.3.2 The ASON/ASTN Framework

ITU-T has followed a formal methodology regarding the integration of the different layers on top of the optical one by first elaborating the requirements for

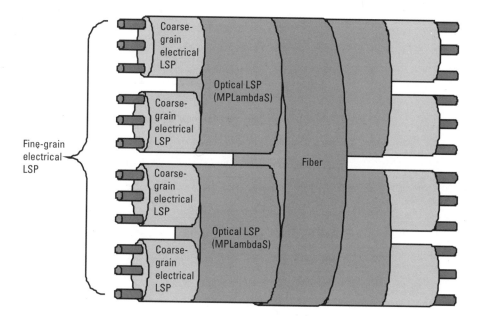

Figure 5.5 GMPLS hierarchy.

such integration proposing a suitable architecture and then continuing with the corresponding detailed design and implementations. This area is covered by ITU-T study groups 13 and 15 working on the direction of ASON/ASTN framework [10, 11]. ASON extends the OTN [12] with an efficient control plane that will make significant savings in the capital and operation expenditures of the operators (see Figure 5.6).

ASON is provided with a control plane capable of provisioning in an automatic and fast way end-to-end optical channel connections as the outcome of a request of any client layer called *user,* such as IP/MPLS, ATM, and SDH, through a UNI, where the "network" is the enhanced OTN. The request initiated through the UNI is further propagated to the ASON network through the use of the NNI. In other words, the UNI is the interface between the control plane of the client layer and the control plane of the enhanced OTN network. Meanwhile, the NNI is the interface between the control planes of the ASON ONEs or ASON subnetworks, either of the same administrative domain, called internal NNI (I-NNI), or of different administrative domains, called exterior NNI or (E-NNI). In addition, ASON promotes the interaction between the management plane and the ASON control plane through a corresponding management interface called network management interface–ASON (NMI-A). NMI-A is capable of triggering and monitoring the request instead of the UNI doing this, and again in this case the latter request is further propagated through the NNIs. The connections that are foreseen in the ASON framework are the *switched optical channels* through the UNI/NNI and the *soft permanent optical channels* through the NMI-A/NNI. Permanent channels are out of scope of the

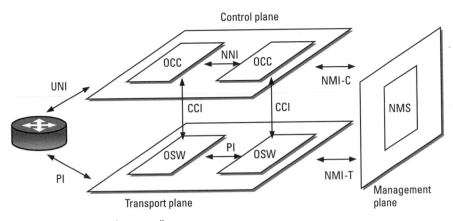

Figure 5.6 ASON architecture (logical view).

ASON/ASTN. ITU-T will try to push the soft permanent schema first, because of its easier and faster deployment as opposed to the switched ones, whose standardization will require much more effort and time.

The network management interface–transport network (NMI-T) is the interface from the management plane (element or network managers) towards the transport network, but is out of the scope of ASON and appears for completeness. The rest of the components are labeled in Figure 5.6. Note that the logical view separating the control plane from the transport plane could be realized as a single merged plane by collapsing the control into the transport plane, as the control plane could be embedded in the optical network elements.

ASTN has generalized the ASON framework and the control/management architecture to cover the SDH layer in a transport-layer independent way, while ASON now focuses on the detailed functional requirements for each transport layer (SDH/WDM). Nevertheless, the ASON/ASTN framework is not bounded yet to any particular signaling protocols, such as the ATM Forum's P-NNI and the IETF's CR-LDP and RSVP-TE extensions. ASON/ASTN deals up to now with the automating provisioning of connectivity in the WDM and SDH layers as a request of a client layer, such as MPLS, but not with MPLS path provisioning itself. Besides connection provisioning and path management, the ASON control plane provides among other functions network and service autodiscovery mechanisms, path selection with routing, directory service, policy control, restoration, and protection.

ASON is being dealt by IETF [13], by different vendors' forums such as the OIF [14] and by multiple European research projects. In Chapter 12, we provide a detailed overview of the IST projects LION [15] and OPTIMIST [16].

5.3.3 The OIF Forum Initiative

OIF was launched in April of 1998 with the objective of fostering the development of a low-cost and scaleable Internet using optical technologies. It is one of the few groups bringing together professionals from the data and optical communities.

In addition to the IETF and ITU-T efforts, the OIF and the Optical Domain Service Interconnect Forum have made one step towards the definition, implementation, and testing of the appropriate optical-UNI signaling messages [17, 18], which will allow the dynamic set up of end-to-end connections between IP routers spanning the optical network. The latter has suspended its operation, as announced on their official site (www.oiforum.com), after completing its main purpose and because most of its work was undertaken by the IETF. UNI version 1.0 successfully demonstrated the interoperability among multiple equipment vendors [19] showing basic path setup functionality. The UNI is currently being enhanced along with work on the NNI.

In any case, the automatic provisioning of end-to-end IP services is still in a testing phase, as all of these proposals are still in an early stage. As standards are still under development, the interoperability of ONEs belonging to different manufactures is an open issue and is not foreseen to be solved in the coming future.

It is obvious that one of the main reasons for these control plane frameworks is the acceleration of the connectivity provisioning. Nevertheless, this is not the case for some of the equipment vendors or telecom operators. Through their sophisticated management systems, they can perform a series of tasks in seconds covering the control plane functionality. In any case, there does not yet seem to be a great need for real-time optical channel provisioning besides grid applications [20], as networks are nowadays overprovisioned and capacity upgrade could be forecasted well in advance with right network planning activities. In addition, interoperability issues among management systems with incompatible or proprietary interfaces were a problem, but the CORBA architecture and open IDL management interfaces of element or network management systems towards the "umbrella" management system have overcome this problem. Finally, features of the control plane such as routing or protection could be integrated in the management systems and the cooperation of control and management plane could bring positive results. The fact that management systems are more mature and ready to offer in the short run services, especially in the optical domain, encourages such cooperation.

5.4 The Management Plane Approach

The term management plane is used in the literature to refer to the set of near real-time mechanisms and algorithms related to the system as a whole and to the OAM. It deals mainly with the procedures related to five functional areas, FCAPS. Path provisioning with routing and QoS support in the configuration management area, and automatic recovery of failures or performance degradations in the fault and performance management areas, are the functions more related to the control plane functionality. Besides these three areas, accounting management is more than necessary for an operator planning to bill services such as automatic path setup and recovery.

Network management functionality mainly exists independently for the IP-electrical world and the WDM-optical world, rather than for the integration of the two worlds. Chapters 3 and 4 have provided enough background on these technology-specific management systems. The next step in network management is aiming toward an integrated management of IP and WDM and other intermediate layers, if any. Such an intertechnology management solution is the main objective this book and will be briefly outlined here.

The proposed management solution is based on an overlay model, as routing is delegated in a two-layer mechanism. In fact, it is performed in an abstracted way in the integrated IP over WDM environment using only edge access points, and the detailed routing is performed inside the IP and WDM domains having all the internal details. Our management solution also identifies a management system for each technology (i.e., IP/MPLS and WDM accompanied by one intertechnology management system). The focus is on the INMS, the MPLS-capable-IP, and WDM management systems, as ATM and SDH and their integration have been dealt extensively in the past [21]. As far as management functional areas, our solution covers configuration, fault, and performance management, although only the first one is dealt with details in the successive chapters of this book. The work done by TMF will be used as a starting point with the appropriate extensions to support the IP technology. In addition, the WDM and IP NMS will be designed and implemented from scratch, following a technology-neutral internal architecture and providing in turn open interfaces towards the vendor-specific WDM and IP EMSs.

The proposed management solution is foreseen only inside one administrative domain for competitive ISPs owning both IP and WDM equipment or for incumbent operator owning multilayer equipment. The main objective of the system is to offer MPLS connectivity services (LSPs), and, as a secondary objective, lambda services (OChs) could be provided to support the required LSPs.

5.5 Summary

Different encapsulation methods have been proposed for the smooth integration of IP over WDM. One such alternative is IP/ATM/SDH/WDM, where IP packets are segmented into ATM cells, which are in turn packet into an SDH frame and transmitted to the optical layer. Another approach considers IP/ATM/WDM, avoiding the SDH layer by means of an ATM cell-based physical layer. More recently, packet over SONET uses encapsulated IP packets that are framed into SDH formats, allowing for a stack like IP/SDH/WDM. Finally, IP over SDL over WDM and IP over GbE over WDM are the most advanced approaches to reach to goal of a simple IP/WDM stack.

In fact, most trends in IP-WDM integration propose the removal of intermediate layers, such as ATM and SDH, and propose the direct interaction of the MPLS-capable IP with the optical layer through a thin adaptation layer. Today's efforts in the telecom industry and the Internet community are characterized by the trend to automate the provisioning of connectivity in terms of OChs in the WDM layer or LSPs in the MPLS layer through the use of signaling protocols. The frameworks working in this direction are the ASON/ASTN towards the first, and MPlambdaS/GMPLS towards both. ASON is pursued among other standardization bodies by the ITU-T SG 13 and 15, while MPLambdaS and GMPLS are related to the IETF technical groups like IP over optical (IPO) and common control and measurement plane (CCAMP). One of the basic reasons for using signaling mechanisms is to speed up circuit provisioning, which usually takes time in the case of OChs. However, this is not the case with the network operators that have developed their own management systems, either WDM or IP/MPLS, that allow hundreds of OChs or LSPs to be set up or torn down each day by means of autorouting and network configuration tools. This book's proposition is based on the latter approach, (i.e., using the management for fast and efficient establishment of MPLS LSP over Optical channels, providing integrated configuration, performance, and fault management functionality, utilizing only the mature features of the control plane as supporting functions).

References

[1] ITU-T Recommendation I.321, "B-ISDN Protocol Reference Model and Its Application," 1991.

[2] Rajagopalan, B., et al., "IP over Optical Networks: A Framework," draft-many-ip-optical-framework-03.txt.

[3] Jain, R., et al., "IP over Optical Networks: A Summary of Issues," draft-osu-ipo-mpls-issues-02.txt.

[4] Awduche, D., and Y. Rekhter, "Multi-Protocol Lambda Switching: Combining MPLS Traffic Engineering Control With Optical Crossconnects," *IEEE Communications Magazine,* March 2001.

[5] "OSPF Extensions in Support of Generalized MPLS," draft-kompella-ospf-gmpls-extensions-02.txt, work in progress, July 2001.

[6] "IS-IS Extensions in Support of Generalized MPLS,"draft-ietf-isis-gmpls-extensions-05.txt, work in progress, November 2001.

[7] OIF contribution #2000.171.0, "LDP Extensions for UNI 1.0," November 2000.

[8] OIF contribution #2000.140.1, "RSVP Extensions for Optical UNI Signalling," August 2000.

[9] "Generalized MPLS–Signalling Functional Description" Draft-ietf-mpls-generalized-signaling-07.txt, work in progress, November 2001.

[10] ITU-T Recommendation G.8070/Y.1302 "Requirements for Automatic Switched Transport Networks (ASTN)," July 2001.

[11] ITU-T Recommendation G.8080/Y.1304 "Architecture for the Automatically Switched Optical Network (ASON)," July 2001.

[12] ITU-T Recommendation G.872, "Architecture of Optical Transport Networks," February 1999.

[13] Aboul-Magd, O., et al., " Automatic Switched Optical Network (ASON) Architecture and Its Related Protocols," draft-ietf-ipo-ason-00.txt, work in progress, July 2001.

[14] Optical Interworking Forum, www.oiforum.com, July 2002.

[15] The IST-LION project, www.telecom.ece.ntua.gr/lion, February 2003.

[16] The IST-OPTIMIST project website, www.ist-optimist.org, December 2002.

[17] User Network Interface 1.0 Signalling Specification, October 2001.

[18] Cao, Y., and Y. Xu, "ODSI Signalling Specification—Optical Gateway Protocol, version 1.0," July 2000.

[19] OIF, "What's New," www.oiforum.com/public/Supercomm.html; Optical UNI Demo White Paper, www.oiforum.com/public/documents/UNIDemoWhitePaper.doc.

[20] Foster, I., and C. Kesselman, *The Grid: Blueprint for a New Computing Infrastructure*, Morgan Kaufmann Publishers, 1998.

[21] Galis, A. (ed.), *Multi-Domain Communication Management Systems*, Boca Raton, FL: CRC Press, 2000.

6

The WINMAN Management System Concept

6.1 Introduction

In the previous chapter, we highlighted several aspects related to the integration of IP over WDM under different points of view; namely the transport plane, the control plane, and the management planes. In this chapter we present what throughout the book is called the WINMAN solution as a set of specifications of a network management system intended to bring a solution for such integration problems. Figure 6.1 depicts the methodology adopted regarding the production of these specifications that lead finally to the functional architecture and functional interfaces of the WINMAN system. The approaches adopted within the different methodological steps are WINMAN specific, and, as such, all them lead to a new solution. Specifically we describe in detail the first three of a sequence of five steps that executed sequentially end with the specification of the WINMAN system architecture and interfaces. These five steps are the following:

1. *Proposal of a business reference model.* The business reference model presents the actors that will make use of the intended system as well as the roles that they may adopt in the interaction with such system. This is the starting point of the whole specification process.

2. *Requirements identification.* In this phase, all requirements are brought together. The requirements are the rules to which the WINMAN solution should adhere. They are categorized, elaborated, and prioritized, until an agreed and feasible set of requirements is achieved.

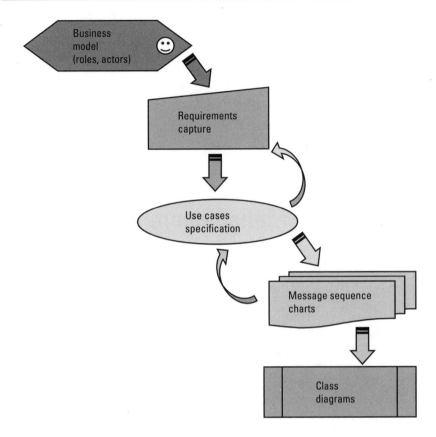

Figure 6.1 The WINMAN methodology.

The approach adopted by WINMAN is to link the actors of the business model with the requirements of the WINMAN solution. In other words, the actors of the business model pose their requirements as external actors to WINMAN, and then the WINMAN system requirements are derived from the needs of the WINMAN actors.

3. *Use cases (UCs) specification.* Following the requirements definition, the next step is to proceed with the definition of appropriate UCs. UCs describe the interaction between the *actors* (both human and nonhuman entities) and the system under design. A UC mainly consists of a number of steps and their variants—effectively scenarios—describing the relevant interactions. This is presented in plain text and complies with a predefined template. Within a UC, there is always a successful scenario description and at least one scenario describing an error condition. Both error and success scenarios have a "most likely" context, thus avoiding description of extreme

occurrences. UCs may describe different parts of the system, but they may also describe the system at different levels.

4. *UML message sequence charts (MSCs) design.* A more detailed examination of a scenario reveals interactions between the actors and the system and within the system itself. These interactions may be specified in a formal manner using MSCs. MSCs consist of sequence diagrams showing interactions among the system entities arranged in chronological order. This step is not covered in the chapter.

5. *UML class diagrams design.* When MSCs are defined for all scenarios, they can be converted into class diagrams. A class diagram describes the types of objects in the system and the various types of relationships that exist amongst them. Furthermore, class diagrams show the attributes and operations of a class and the constraints that apply to the way the objects are connected. The approach that has been adopted consists of converting the messages in the MSCs to class methods and the entities in the MCS to the classes. The attributes of each class can be identified from the messages exchanged (the MSC message attributes—contained within brackets—become class attributes). No further detail on this step is described here.

6.2 Business Reference Model

The goal of this section is to identify the entities that will interact with the WINMAN system, as well as their roles and what they are expecting to get from the system. After that, the business roles that the system will address are specified, giving the necessary means to define the high-level architecture.

6.2.1 Business Roles and Actors

Business roles represent activities that businesses can engage in. By focusing on roles rather than organizations, a more flexible business model can be achieved, as organizations can change roles dynamically or even adopt more than one role.

The following types of business roles were identified as having some kind of interaction, either directly or indirectly, with the WINMAN system:

1. Customers and end users:

   • Customers are legal persons, either humans or companies, that have contracts with service providers about the right to use telecommunications services and the obligations to pay for this right and the usage of these services according to the tariffs.

- End users are entities that interact with the service providers to obtain the effect of the service. End users may be humans or an automated piece of application software.

- This is their only business role—consuming services—and not trying to make a profit out of them.

2. Value-added service providers (VASPs), whose role is oriented towards customer management and adding value. Customers buy services from VASPs, which act as retailers of telco services, providing services other than connectivity.

3. Network service providers play the role of both network provider and network management provider. Their role is to support VASPs to provide their services, acting basically as managed connectivity providers. They give a management interface to VASPs, which enables them to request connections between arbitrary end points in the global network.

Note that these business roles play the role of user and provider towards other business roles For example, a VASP provides services to customers—the provider role—but uses services from network service providers—now playing the user role. Several business relationships can be established among the different roles.

Several roles can be performed by a single business entity (the actor or player or stakeholder) at the same time. For example, a player in the network service provider role can also be in the role of a VASP. This is the case for companies who own their own transport network and want to make revenue in the new markets (e.g., an incumbent telco operator).

The following actors can be defined:

- Incumbent/traditional telecommunication providers that used to be called PNOs: these players have an incumbent position in the market, as they have extended fixed network infrastructure, both in the local loop (copper) and in the backbone all over the country. They can play multiple roles in once, such as the roles of network service provider, VASP and Internet service provider (ISP).

- Competitive/new entrant telecommunications providers: these providers are new entrants following the market deregulation and are competing with the incumbents. They usually do not have extended network infrastructure in the local loop and build their backbone on top of the incumbent ones. They use the local loop either by the unbundling deregulation procedure or by owning licenses in the wireless local loop. They gradually build their own backbone network. Examples of such carriers are the ISPs, who extend their activities into telephony and other services.

6.2.2 Business Case

Operators wish to build carrier-scale multiservice IP networks to provide a wealth of new multimedia services to their customers. All of these services have to be cost effectively, rapidly and reliably deployed to millions of customers.

In parallel to the deployment of the network, operators are faced with the challenge of installing their respective OSSs. The next paragraphs describe the competitive advantages that the availability of a system based on the WINMAN solution will provide to the network service providers.

6.2.2.1 Service Provisioning

To date, ISPs have opted to offer one-size-fits-all, flat-rate plans that allow customers unlimited use of standard services. The ability to offer premium-grade services such as guaranteed bandwidth, voice over IP, and VPN are dependent on the ability to define personalized user policies that map users to the class of service to which they are subscribed. WINMAN will make policy-based IP connectivity service provisioning a reality, thus empowering service providers to quickly bring new differentiated services to market. In addition, WINMAN will pave the way for flowthrough automation in conjunction with other OSS applications, by means of standardized interfaces. WINMAN will consider system level specification (SLS) parameters derived from SLAs, thus fulfilling the required QoS.

6.2.2.2 Network Provisioning

The network provisioning as well as the network infrastructure itself has become increasingly complex. The manual configuration of service *pipes* on the network is costly, time consuming, and error prone. As the demand for timely delivery of more innovative services increases, the requirement to automate the provisioning process is stronger than ever. Automating the provisioning process allows service providers to scale their operations and improve the quality of end-to-end deployment for new customers.

The main functions in the network provisioning process are:

- *The provisioning of end-to-end IP paths over lightpaths using MPLS technology.* In this context the WINMAN system will be capable of calculating, designing, and creating MPLS LSPs over the corresponding lightpaths in the optical domain. For completeness as well as for backwards compatibility, the provisioning of IP plain connectivity (connectionless mode) is also considered.

- *Support of traffic and QoS parameters for MPLS LSPs derived from SLAs.*
  Policies and other requirements will be applied in the path-provisioning
  request.

Secondary functions supporting these are intended to discover network
resources, network topology, and maintenance of an inventory of all network
resources with their status and their hierarchical relationship; to notify the service management about service status; and to provide updates to the fault and
performance management layers reflecting the changes in the network configuration and new services.

WINMAN offers an integrated and automated provisioning solution for
network service providers that offer IP-based network services (i.e., IP-VPN,
VoIP, and MoIP), transparently and independently from the underlying transport network (i.e., ATM, SDH, or WDM).

6.2.2.3   Fault Management

In today's service provider environment, it is no longer acceptable to simply
monitor for a network's device-specific fault events. In a world where SLAs
impose penalties for network down time, network events must be correlated
with affected customers. Additionally, service providers can differentiate VPN
services by offering customers a "window" into the events associated with their
VPNs.

WINMAN offers a service level alarm monitoring and diagnostics tool
that provides network fault monitoring, network trouble isolation, and network
recovery. WINMAN is designed to help operators focus on important network
events by offering a combination of alarm processing rules, filtering, and customizable alarm viewing. It will offer Web-based interfaces, which allow service
providers to select specific views of events and availability reporting associated
with particular services.

The main functions in the fault management area are:

- *Report faults in the IP or optical domain in an intelligent and integrated
  way.* Reporting of primary faults, either in the electrical or optical
  domain, should be supported after the corresponding filtering, analysis,
  and correlation of the multiple alarms that are propagated in case of a
  single fault. The report should include all of the attributes of the antici-
  pated alarms together with the list of affected LSPs.
- *Recover faults in the IP or optical domain in an integrated way.* In case the
  automatic protection switching is not applicable in any of the domains
  or it fails partially or completely, an automatic fault restoration mechanism should apply to try to restore all the affected LSPs or IP plain

connectivity. This automatic procedure is triggered by the management system after the integrated analysis and correlation of the propagated alarms.

Secondary functions supporting these are intended to interface with the network topology database; keep an inventory of user-defined correlation rules; maintain a fault topology database that contains the alarm status of the network resources and services; design management rules for the multilayer survivability aspects (i.e., avoidance of conflicting actions, protection and restoration, and service differentiation), and handle the service requests from service management systems asking for the fault status of the connectivity services.

6.2.2.4 Performance Management

Service providers face the challenge of the ability to measure, not just network element throughput, but also user experiences through their networks.

WINMAN will monitor key SLS metrics such as response time, availability, jitter (interpacket delay variance), connect time, packet loss, and application performance. Also important is the ability to measure performance within a particular customer's VPN.

The main function is to monitor, filter, and report performance data. The system shall monitor the basic traffic and QoS network parameters of the LSPs and report service degradations in case of performance gauges or counters threshold crossings.

Secondary functions are those intended to set threshold crossing alerts on the available route capacity between any two-service locations for all provided transport services/facilities or on the equipment capacity; handle notifications of capacity threshold crossings; obtain periodic and on-demand reports of the monitored capacity (traffic load); and use the obtained traffic monitoring data for identifying hot spots in the network and take actions to prevent network congestion.

To sum up, WINMAN brings an IP over WDM connectivity service provisioning system, able to integrate and correlate fault and performance information across different technological domains—IP and WDM at least and ATM and SDH whenever they are used. This is why we use the term interdomain management system to refer to the WINMAN system throughout the book.

6.2.3 Business Processes Addressed by WINMAN

TMF has launched significant activity to capture the needs of network operators and service providers and thus enable the "end-to-end process automation of telecommunications and data services operations processes" [1]. The telecom operations map (TOM) is the framework for accomplishing this mission. The TOM defines the business processes and their interactions used by service providers in the customer, service, and network management areas.

The TOM decomposition of business processes has been adopted by WINMAN, followed by the proper adaptation and customization. Also, the requirements capture following in the next section will be based in the TOM business process decomposition.

6.2.4   High-Level Management Architecture

In this context the WINMAN scope can be defined through a model following the TMN architecture. Specifically, WINMAN focuses on the network management layer of the TMN OSFs pyramid. The reference points for interactions with the outer world are the northbound interface towards other service management function systems (such as those providing VPN-IP, VoIP, and MoIP) and the southbound to the element management function systems (such as those managing IP, ATM, SDH, and WDM network elements). Although the WINMAN solution is valid for any of the earlier mentioned technology, only IP and WDM will be considered in detail for implementation purposes. Finally, a presentation function system or workstation function in TMN terms is provided to the WINMAN operator as a mean of controlling and monitoring the WINMAN functionality. Although WINMAN is essentially focused on the network layer, light service layer functionality is also provided for testing and demonstration purposes. This is summarized in Figure 6.2.

6.3   Requirements Specification

The approach adopted in WINMAN is to link the actors of the business model with the requirements of the WINMAN solution. In other words, the actors of the business model pose their requirements as external actors to WINMAN, and then the WINMAN system requirements will derive from the needs of the WINMAN actors. The WINMAN actors are also present in the UCs, which will lead to the WINMAN system functionality and functional architecture. The UCs are coupled with the appropriate set of requirements. Figure 6.3 depicts the WINMAN actors. The WINMAN requirements will be captured according to needs of these four external actors—the service management system, the WINMAN operator, and the IP and WDM element management systems. For this reason, the system will be treated like a black box, where the internal structure and system components are irrelevant at this stage.

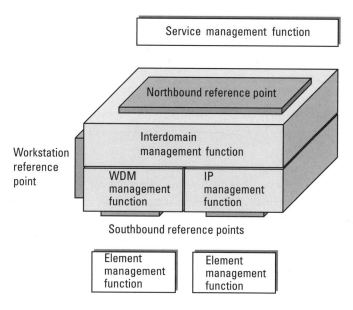

Figure 6.2 The WINMAN high-level management architecture.

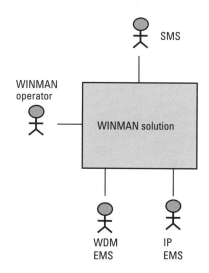

Figure 6.3 Actors identified in the WINMAN business model.

6.3.1 Requirements Classification

The system requirements are divided into several groups, each one of them addressing a different business case. Each requirement is identified with a unique identifier (ID) so it can be easily addressed in any other document.

Moreover, each requirement is prioritized (different levels of priorities are defined), classified as mandatory (for requirements that must be implemented) or desirable (for requirements that provide some value added to the system but do not need to be implemented for the system unless we want a fully functional system).

The following sections describe briefly the groups of requirements. Under another point of view, the requirements are split in two main categories:

1. Functional requirements, describing the requirements of the external actors of the WINMAN solution with respect to configuration, fault, and performance management functionality;

2. Nonfunctional requirements, such as interoperability, implementation, and integration specific requirements among others.

As described in Section 6.2.3, the requirements identification is based on the TMF TOM approach. This means that the requirements are also categorized per functionality belonging to business processes defined in TOM.

In this section, the WINMAN solution requirements are presented in detail according to the TOM business processes. We start giving an overview on the WINMAN data repositories, and we end with a set of testing requirements.

6.3.1.1 Data Repositories

The WINMAN solution shall consist of three types of databases used for network configuration purposes at different levels and another two types for fault and performance data:

- *Network inventory or physical database.* The network inventory database is managed by the network inventory management (NIM) subsystem (according to the TOM processes) and maintains the physical inventory of the network resources. The database repository resembles a hierarchical structure, representing the equipment in the network. Minimally, the following attributes are found for each network resource of the network inventory database: cards, ports, fiber conduits, and location.

- *Logical database.* The logical database is managed by the network provisioning (NP) system. The logical database maintains the logical inventory of the network resources and consists of entries representing logical paths, which are supported by the physical resources of the network. Each path is specified by an entry in the database having at least the following attributes: protection of the path, available bandwidth, and network quality expressed in terms of delay and delay variation.

- *Connectivity database.* The connectivity database is managed by the NP subsystem. All connections in the network are stored with their attributes in one entry in the connectivity database. If a particular connection is released, the corresponding entry will be removed from the connectivity database. A connection minimally has the following attributes: the connection status (pending or in effect), the logical route through the network, the connection bandwidth, and the connection quality (e.g., delay or jitter).

- *Fault topology database.* The fault topology database is managed by the network management and restoration subsystem. This database has an entry for each root cause problem within the network. The root cause has at least the following attributes: fault cause, fault severity, and fault location.

- *Performance management database.* The performance management database is managed by the network data management subsystem. This database stores the quality levels in each of the logical paths within the network.

6.3.1.2 NP

The aim of the NP process is the configuration and installation of the logical and physical network. The process ensures that network capacity is ready for provisioning and maintenance of services. It carries out the NP, as required to fulfill specific service requests, network and information technology additions, changes, deletions, and configuration changes to address network problems. The NP process administers the logical network and interfaces with the network inventory management process. These are some of the requirements that the WINMAN system must fulfill:

- The WINMAN system must be able to receive, process, and implement service requests, changing, if necessary, the configuration of any technological layer of the managed network.

- The WINMAN system must guarantee QoS.

- It must be possible for the user of the service to change its characteristics or to remove it, after the request is made.

- The WINMAN system must support scheduled requests (requests made for connection services that should be available in the future for a certain amount of time).

6.3.1.3   Network Inventory Management

The aim of the network inventory management (NIM) process is to install and administer the physical infrastructure of the network. The responsibilities of the process concern the installation and acceptance of the equipment, the physical configuration of the network, and management of the spare parts, the part return/repair subprocesses, and equipment software upgrades.

The system should be able to build an inventory with all of the resources and equipment of the network, and that inventory should keep track of all of the network topology changes.

6.3.1.4   Network Maintenance and Restoration

The aim of the network maintenance and restoration (NM&R) process is to maintain the network, eliminating or minimizing the impact of problems caused by the infrastructure; to restore or repair customer-affecting troubles quickly, and to identify events in the network prior to those events becoming customer-affecting problems. This process concerns maintenance of the operational quality of the network according to the required network performance goals.

The following are some of the most important requirements defined in this section:

- The system must receive notifications about network events from the element managers.
- If there is a fault in the network that compromises a service, the user who requested the service must be informed.
- The WINMAN system should be able to provide alarm-filtering tools to the users.
- The system must be able to perform alarm correlation to find the cause of the faults in the network.
- The system should reroute affected services to restore as soon as possible the affected services.

6.3.1.5   Network Data Management

The aim of the network data management process is to ensure that network performance goals are tracked and notifications are provided when they are not met. The process must provide sufficient and relevant information to verify compliance to SLSs and QoS levels. Some of the requirements are the following:

- The WINMAN system should monitor the QoS parameters of the active connectivity services.

- The system must be able to generate alarms when previously established QoS thresholds are crossed and the fulfillment of the connectivity services, as agreed with the users, is in danger.

- The system should trigger, if necessary, network maintenance and restoration processes to prevent the nonfulfillment of QoS guarantees.

6.3.1.6   Test Requirements

The purpose of this section is to define the service layer requirements to test the WINMAN solution. This is performed in order to test the WINMAN system, although the service layer is not considered the focus of WINMAN.

- *Service configuration.* The aim is to provide service configuration or reconfiguration, including connection management activities, within the time frame required to meet the ever-increasing service intervals in support of on-time delivery to customers. The process covers the installation and configuration of service for specific customers. The reconfiguration of service after the initial service installation is also supported.

- *Service quality management.* The aim of the service quality management process is to provide service specific monitoring and to ensure that service performance meets or exceeds committed levels. The process supports monitoring of the service quality and cost on a service basis in order to determine whether service levels and costs are met consistently, or if there are problems or improvements that can be made, and whether the sale and use of the service is tracking the forecasts.

- *Network development and planning.* The aim of the network development and planning process is to design, develop, and deploy a low-cost network and information technology infrastructure that meets the requirements of services provided on that network. The process encompasses the development and acceptance of network and information technology infrastructure strategies. It also describes the standard network configurations primarily for operational use and defines the rules for networks (e.g., planning and installation). It carries out the design of the network capabilities to meet a specific service need at the desired cost and the design, deployment, and introduction of new technologies to support new services, features, or enhancements. Proper installation, monitoring, controlling, billing, and meeting the forecasted demand of the network must be ensured. The process also supports cases of unforecasted demands.

6.3.2 Nonfunctional Requirements

The most important nonfunctional requirements are the following:

- The WINMAN design and implementation shall be based on specific standards, so as to be able to interoperate with other systems over well-defined interfaces.
- The WINMAN solution should be tested in real conditions before being put into operation.

Interactions between WINMAN components shall be minimized in order to minimize the bandwidth required by the system.

6.4 WINMAN UCs

After the requirements capture, the next step in defining the system's architecture consists of the identification and description of appropriate UCs that illustrate the basic functionality of the target system. The architecture can only be fully defined when the MSCs and class diagrams are defined. Appropriate terminology has been used for building the UCs. Relevant references include (but are not limited to) [2–6].

6.4.1 Identification and Classification of UCs

Table 6.1 depicts the names and allocation of UCs into appropriate categories. The horizontal classification of UCs relates to the functional grouping of UCs and is compliant to [1] in terms of the terminology used. However, the policies provision and scheduler groups of UCs are WINMAN specific. The vertical classification of UCs is according to the level of abstraction and related context and modularity of the UCs.

An identifier is allocated to each UC (in parenthesis). The scheme adopted is XYZ, where X denotes the horizontal classification (functional grouping), Y denotes the vertical classification (level), and Z is a sequential number.

6.4.2 UCs Overview

This section presents an overview of all identified UCs. Critical aspects regarding their context are briefly discussed, while UCs interrelations are depicted in UML diagrams, according to the previous classification. Due to the nature of interactions, some UCs appears in multiple diagrams. A short description of the goal and context of each UC is also provided.

Table 6.1
Classification of Use Cases

	Summary Level (S)	User Level (U)	Atomic Level (A)
Policies provision (P)	Provide policies (PS1); Provide policies report (PS2)	Add policies (PU1); Modify policies (PU2); Activate policies (PU3); Execute policies (PU4); Remove policies (PU5)	Check policy (PA1); Find policy (PA2)
Scheduler related (S)		Schedule task (SU1); Delete task (SU2); Modify task (SU3)	
Network provision (N)	Provide integrated connectivity service (ICS)-SMS (NS1); Provide ICS-WO (NS2); Modify ICS-SMS (NS3); Modify ICS-WO (NS4); Delete ICS-SMS (NS5); Delete ICS-WO (NS6)	Verify connectivity service (CS) request (NU1); Create CS-SMS (NU2); Create CS-WO (NU3); Implement CS (NU4); Activate CS (NU5); Deactivate CS (NU6); Release CS (NU7); Remove CS (NU8)	Compute list of IP routes (NA1); Compute IP paths under constraints (NA2); Compute list of optical trails (NA3)
Network inventory management (I)	Register EMS (IS1); Unregister EMS (IS2); Notify network change (IS3)	Configure network (IU1)	Add IP node (IA1); Delete IP node (IA2); Add WDM node (IA3); Delete WDM node (IA4); Add WDM link (IA5); Delete WDM link (IA6); Modify WDM link (IA7); Add EMS (IA8); Delete EMS (IA9)
Network maintenance and restoration (M)	Recover network (MS1); Revert network (MS2); Open alarm (MS3); Clear alarm (MS4); Provide alarm reports (MS5)	Identify affected connections (MU1); Identify root cause (MU2); Synchronize alarms (MU3); Acknowledge alarm (MU4)	
Network data management (D)	Process performance measurements (DS1); Provide PM report (DS2)	Calculate aggregations and statistics (DU1)	

6.4.2.1   Policy Provision

Figure 6.4 depicts the policy provision UCs considered in WINMAN.

- *Provide policies UC.* The purpose of providing policies at the summary level is to allow the WINMAN operator to add/modify/execute/delete a policy in/from the WINMAN system.

- *Provide policy report UC.* This UC at the user level creates reports regarding the policies that exist in the system. The nature of the reports is generic, ranging from simple statistics to presentation of more complex policy-related data at summary level.

- *Add policies and remove policies UCs.* The purpose of these UCs at the user level is to install a new policy in the WINMAN system, leaving it ready to be activated, or to remove it from the repository, respectively.

- *Modify policies UC.* The WINMAN operator modifies an existing policy, at user level.

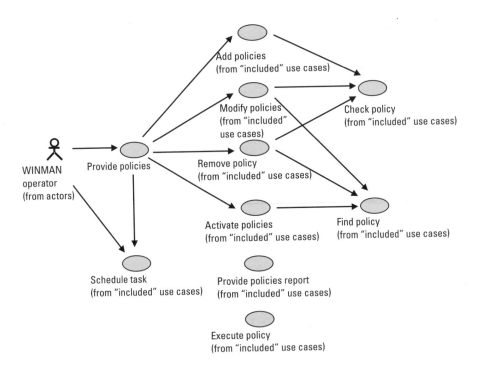

Figure 6.4 Policy provision UCs.

- *Activate policies UC.* The purpose of activating policies at the user level is to instantiate a certain set of policies in the appropriate system components, leaving it ready to be used by the execute policies UC.

- *Execute policies UC.* The purpose of execute policies at the user level is to carry out actions according to an instantiated policy. The policy conditions are checked and the relevant actions are executed. This UC may be called from any of the NP, NIM, NM&R, and NDM UCs.

- *Check policy UC.* The check policy at the atomic level is intended to determine both the policy validity in itself and whether it is in conflict with all other existing policies. The principal mechanisms driving this check will be dictated by rules specifically designed for such a purpose. These rules are called metapolicies.

- *Find policy UC.* The purpose of this UC at atomic level is to find a given policy in the repository.

6.4.2.2 Scheduling

The scheduler is a part of functionality in the WINMAN system to perform a variety of tasks on such areas as (but not limited to):

- Connection management (implementing/activating a network connection);

- Configuration management (periodically making changes in the network);

- Fault management (providing fault management summary reports once or on a certain time base, or changing correlation rules);

- Performance management (providing time-based reports or changing thresholds);

- System administration (performing backups);

- Policy management (executing policies).

The scheduler keeps a list of tasks to be scheduled. These tasks are in principle performed in a chronological fashion but also take into account the task priority. However, policies that apply to the scheduler can override the execution order.

The task to be scheduled can have a fixed date and time but can also have a periodical schedule. The task owner is entitled to request the scheduling of a task. Two possible owners are distinguished; a task is owned either by the WINMAN operator or by the connection management UCs (provide ICS/delete ICS/modify ICS).

- *Schedule task UC.* A task is added to the task list of the scheduler.

- *Delete task UC.* A task is deleted from the task list of the scheduler.

- *Modify task UC.* A task is modified in the task list of the scheduler.

Figure 6.5 depicts the UCs that implement the WINMAN scheduling functionality.

6.4.2.3   NP

As far as NP is concerned, we consider three types of operation—automatic, semiautomatic, and manual. In automatic mode, the user (SMS or WINMAN operator) just requests a connection between two or more end points with a set of constraints. The system calculates all paths that connect the end points according to the constraints and selects the best one. In semiautomatic mode, the system works just like in automatic mode, but instead of choosing the best path by itself, it lets the WINMAN operator do it. In the manual mode of operation the WINMAN operator designs the path. WINMAN then validates the path checking if it really connects the end points and whether it can guarantee the bandwidth and QoS-requested parameters.

WINMAN Support for Connectivity

WINMAN supports two types of connectivity services:

1. *IP connectivity relating to the setting up and maintenance of end-to-end IP paths.* This type of connectivity service is available in all operational modes as specified earlier and can be managed by both the WINMAN operator and the service management system. IP connectivity is

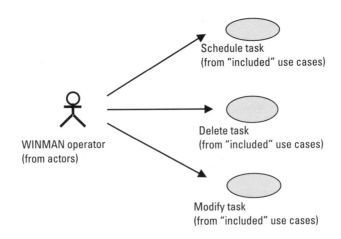

Figure 6.5 Scheduling UCs.

considered to be the main focus of the WINMAN solution and includes the set up of connections in the WDM layer to accommodate overlying IP connections in a transparent manner.

2. *Optical connectivity relating to the setting up and maintenance of optical trails by the WINMAN operator in the manual mode only.* This feature is valuable in cases where it is desirable to set up manually connections in the WDM layer (e.g., optimizing network capacity/performance or dealing with optical element malfunctions).

When a connectivity service of the first category is requested to the WINMAN system, we refer to it as an ICS. On the other hand, when the connectivity request belongs to the second category, we term it simply as a CS. The term CS is also extended to IP connectivity directly requested from the IP layer and not through the integrated management layer. In other words, a CS is a single domain connectivity request, whereas an ICS might involve both the IP and the WDM domains.

Integrated Connectivity Service and the Network Resources States

Each step performed by the WINMAN system in the *provide ICS, delete ICS,* and *modify ICS* UCs brings the ICS and the network resources into another state. Figure 6.6 provides an overview of the different states of an ICS. The ovals represent the states and the labeled arrows represent the processes to be executed to get from one state to another.

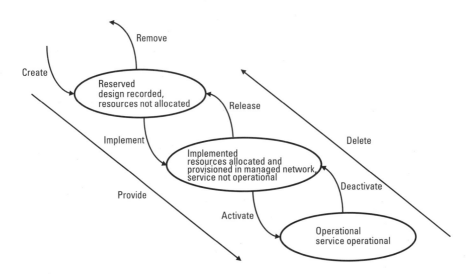

Figure 6.6 The ICS state diagram.

The following ICS states are distinguished:

- *Reserved.* The ICS is created, which means that the network connections are designed and the required network resources are determined and provisioned (parameters have their values). The design of the ICS is recorded in the WINMAN system. The network resources are reserved in the interval(s) in which the ICS becomes operational. The WINMAN system is ready to implement the ICS.

- *Implemented.* The ICS is implemented, which means that the network resources are allocated and thus cannot be used by another ICS. The managed network is accessed for the allocation of the physical network resources and the testing of the network connection(s). The WINMAN system is ready to activate the ICS.

- *Operational.* The ICS is in operational state, which means that the network connections are available to the end users, and maintenance processes have started for this ICS.

- *Provide ICS UC.* This summary-level UC encompasses the functionality involved in providing a connection between end points and uses several user-level UCs to fulfill its goals. These include all of the steps from connection request verification up to the activation of the connection (bringing it into the operational state). Provide ICS is capable of connecting IP end points, including the setting up of connections in the WDM layer in a transparent way. Figure 6.7 depicts the UCs involved in providing an ICS.

  Two separate UCs are defined for supporting the provision of IP connectivity services. The first is to provide ICS by the SMS, whereby the service management system requests from the WINMAN system to provide an ICS. The second is to provide ICS by the WINMAN operator. Here, the WINMAN operator requests the WINMAN system to provide an ICS. The following processes (steps in the UCs) are distinguished when providing the ICS:

  1. *Verify the ICS request.* This step describes the WINMAN system responsibilities for the verification and validation of the ICS request to the WINMAN system. The WINMAN system verifies the correctness of all of the attributes of the request and determines the validity of the request itself (if applicable) by applying policy rules to this request.

  2. *Create the ICS.* This step describes the WINMAN system responsibilities for the design of the network connections that supports the ICS and the determination of the required network resources

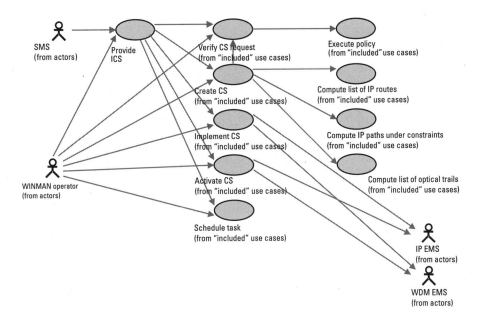

Figure 6.7 Provide ICS UCs.

for these network connections. The design is filed in the WINMAN system. The required resources are reserved for the timeslots in which the ICS will be operational. The ICS state becomes *reserved*.

3. *Implement the ICS.* This step describes the WINMAN system responsibilities for the implementation of the network connections. The implementation starts when the ICS is at the point of becoming operational (scheduling). Then the WINMAN system allocates the required network resources and changes their usage state from *idle* to *active* or *busy*. So far, the managed network has not been involved in the implementation. Then, the WINMAN system allocates the physical resources in the managed network through the IP or WDM EMSs. Subsequently, the parameters of the network resources are provisioned, and the network connection is set up and tested. The ICS state becomes *implemented*.

4. *Activate the ICS.* This step describes the WINMAN system responsibilities for the activation of the ICS. This step involves the interaction of the WINMAN system with the managed network, which leads to the operational state of the ICS. The ICS state becomes *operational*.

- *Modify ICS UC.* The modify integrated connectivity service UC is used when one or more service parameters need to be changed. The WINMAN operator and the SMS can request the change of service characteristics. Figure 6.8 depicts the UCs involved in modifying a connectivity service within WINMAN.

Two separate UCs are defined for supporting the modification of connectivity services:

1. *Modify ICS by the SMS.* The main flow described in this UC is a modification of the service in automatic mode. The modification is done after a single request from the SMS. A step-by-step modification, driven by several requests from the SMS, is described as an alternative flow.

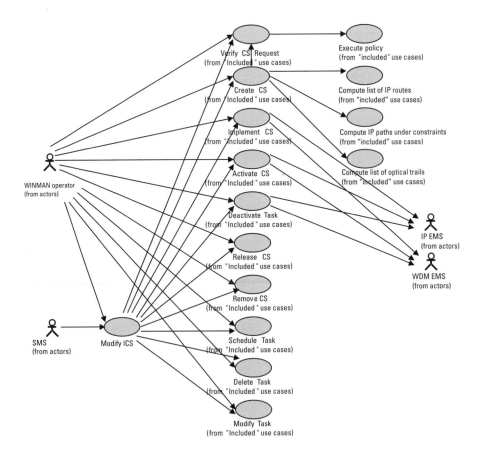

Figure 6.8 Modify ICS UC.

2. *Modify ICS by the WINMAN operator.* This UC describes how the WINMAN operator can modify an integrated connectivity service.

The inherent flexibility in the IP layer allows for the modification of a service without service outages. This means that with the ICS in active state, the WINMAN system can implement the changes on the network.

The following modifications of a service have been identified:

1. *Change of the SLSs, such as, but not limited to, bandwidth, reliability (availability or protection level), delay, jitter, or packet loss.* This modification may not require changes in the managed network, if the existing IP path that supports the service meets the new QoS parameters.

2. *Add or remove a leg to a point-to-multipoint or multipoint-to-multipoint service.* This functionality will not be included if theses services are not included in WINMAN.

3. *Change the scheduling (initial or final time) of the service.* The SMS or the WINMAN operator can trigger the UC. In the first case, the process will be automatic; that is, the SMS requests WINMAN to modify the service parameters, and WINMAN carries out the needed actions. In the other case, the operator can have more control over the process. The modify ICS request may or may not cause changes in the network connectivity. According to the requested modification, the service may have to be rerouted through a new path that can satisfy the new SLSs. During the modification, the service may become unavailable for some time.

- *Delete ICS UC.* This UC describes the opposite of the provide ICS UC. This is the summary UC described in the most general terms. Figure 6.9 depicts the UCs involved in deleting an ICS.

Two separate UCs are defined for supporting the deletion of ICSs:

1. *Delete ICS by the service management system (SMS).* The SMS requests from the WINMAN system the deletion of an ICS. It is assumed that an ICS is deactivated and released at the end of a session.

2. *Delete ICS by the WINMAN operator.* The operator requests from the WINMAN system the deletion of an ICS. It is assumed that an ICS is deactivated and released at the end of a session.

The following processes are distinguished when deleting the ICS:

1. *Verify ICS request.* This step describes the WINMAN system responsibilities for the verification and validation of the ICS

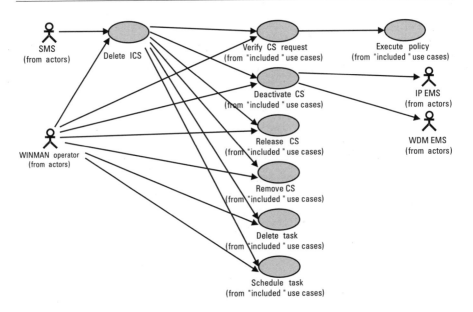

Figure 6.9 Delete ICS UC.

request to the WINMAN system. The WINMAN system verifies the correctness of all of the attributes of the request and determines the validity of the request itself (if applicable) by applying policy rules to this request.

2. *Deactivate ICS.* This step describes the WINMAN system responsibilities for the deactivation of the ICS. This step involves accessing the managed network through the IP and WDM EMSs for the deactivation of the ICS. After that, the WINMAN system updates the state of the ICS. The physical network resources are still allocated and the connections are still set up but are not available to the end user. The ICS state becomes *implemented.*

3. *Release ICS.* This step describes the WINMAN system responsibilities for the release of the ICS, and it involves the interaction of the WINMAN system with the managed network through the IP and WDM NMSs for the release of the physical resources in the network. The WINMAN system also releases the logical network resources, which makes them available for other ICSs. The state of the involved network resources becomes idle (if no other ICS currently makes use of the resource) or active (if another ICS currently makes use of the resource). The design of the ICS remains filed in the WINMAN system; it may become operational in another time interval. The ICS state becomes *reserved.*

4. *Remove ICS.* This step describes the WINMAN system responsibilities for the removal of the ICS from the list of ICSs. The record of the ICS is removed, and the ICS does not exist anymore.

    - *Verify CS request UC.* The verify CS request is considering the verification of CS requests from both the WINMAN operator and the SMS. This two-step method (verify semantics and validate against policies) could be practically applied for any type of request verification as long as there is an appropriate algorithm (policy) to handle different sets of attributes.
    - Create CS UC:
    1. *Create CS by the SMS.* This UC is called from a summary level UC for the design of a CS between two or more end points. The system has to find the best path that fulfills the requirements proposed and make the necessary reservations of resources on the data storage facilities. No changes on the network are made at this time. The UC is prepared to support the modes of operation automatic and semiautomatic. In the automatic and semiautomatic mode, the best path found is returned automatically. In the manual mode, the WINMAN operator is prompted to select one of the paths found.
    2. *Create CS by the WINMAN operator.* This UC is used directly from the WINMAN operator to create connectivity between two or more end points either at the IP or the WDM layer (it is not possible to have interlayer connections). The WINMAN operator gives a list of the links he or she wants to use to connect the end points, and the system has to check if the path fulfills the requirements proposed and make the necessary reservations of resources. No changes on the network are made at this time.

- *Implement CS and release CS UCs.* The implement CS UC describes the process of implementing a CS. The CS under consideration (if in the IP domain) is expected to be in the reserved state and will change to the implemented state, if this UC is successful. The scheduler invokes the implement CS UC when the time for implementing the CS is reached. WINMAN requests the involved EMSs to establish and test the connectivity. The release CS describes the process of releasing a service. The service goes from the *implemented* to the *reserved* state. Note that in the *reserved* state the design of the service is kept. Resources for a future activation of the service are still reserved, but until then they can be used by other services.

- *Activate CS and deactivate CS UCs.* These UCs incorporate the functionality for activating an already implemented CS or deactivating an active one. The WINMAN operator or the SMS does not invoke either UC directly, but via the provision ICS or modify ICS request. The scheduler triggers indirectly, via the relevant summary level UCs, the activate CS UC, or the deactivate CS UC when the time for doing so is reached.

- *Remove CS UC.* This UC is used by the delete ICS UC, or the WINMAN operator to eliminate all of the reservations of resources for the CS on the storage facilities. The removal of reservations is made only on the WINMAN storage facilities.

- *Compute list of IP routes UC.* This UC is constructed to be called from other UCs (create CS, for example). It accepts a list of two end points and calculates all the IP routes (combinations of IP links) that connect one end point to the other. The algorithm searches the IP network elements database for one of the end points (routers) of the connection. That end point will be considered the root. Starting at the root, the algorithm must analyze all of the branches and discover which combinations of branches can lead to the other end point.

- *Compute IP paths under constraints UC.* This UC is constructed to be called from other UCs (create CS, for example). This UC accepts a list of IP routes and a list of constraints, expressed as boundaries on a set of parameters. By comparing all of the free resources on the IP links used by all the routes with the list of constraints, it discovers which routes can fulfill the constraints. An IP route is composed of IP links. All of the IP links of each route will be analyzed and checked against the relevant set of parameters. The links that cannot fulfill the requirements are marked, and all of the routes that use them are rejected.

- *Compute list of optical trails UC.* This UC is constructed to be called from others UCs (create CS, for example). This UC accepts a list of two end points and calculates all possible optical trails that connect one end point to the other. The process is very similar to the compute list of IP routes UC. The algorithm searches on the WDM network elements database for one of the end points of the trail. That end point will be considered the root. Starting at the root, the algorithm will have to analyze all of the branches and discover which combinations of branches can lead to the other end point.

6.4.2.4 NIM

The NIM UCs incorporate all of the functionality necessary for managing the WINMAN inventory that contains the configuration of the network managed by WINMAN. Figure 6.10 depicts the UCs related to NIM.

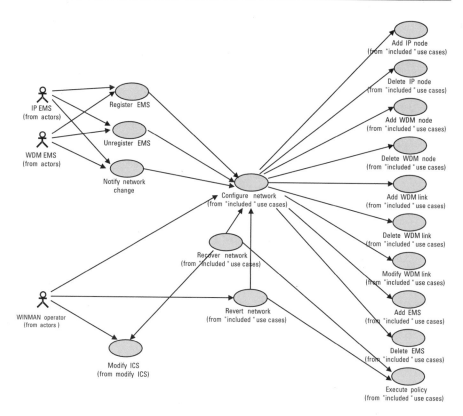

Figure 6.10 NIM UCs.

- *Register EMS UC.* This UC registers a new EMS to the WINMAN system, thus configuring and activating a communication link towards this EMS within the WINMAN southbound interface. Therefore, it constitutes an autodiscovery process of the new EMS.

- *Unregister EMS UC.* This UC can unregister an EMS to the WINMAN system, thus dropping a communication link towards this EMS within the WINMAN southbound interface. It constitutes a counterpart of the register EMS UC.

- *Notify network change UC.* This UC handles notifications regarding changes in network elements, received from the WINMAN southbound interface (i.e., the EMSs managed by WINMAN). It requires that a communication link towards the EMS is active for the exchange of messages at the WINMAN southbound interface.

- *Configure network UC.* The configure network UC is used when one or more network inventory parameters must be manually changed (including addition and removal of network elements) by the WINMAN operator. This interaction is done through the appropriate graphical user interface (GUI) representations of the current network configuration map.

- *Add IP node and delete IP node UCs.* The add IP node and the delete IP node UCs are used by the configure network UC when new IP nodes need to be added or deleted to or from the WINMAN system, respectively.

- *Add WDM node and delete WDM node UCs.* These UCs create or delete a WDM node by adding or removing it in the WINMAN inventory.

- *Add WDM link and delete WDM link UCs.* These UCs create or delete a WDM link entity in the WINMAN inventory.

- *Modify WDM link UC.* This UC modifies the characteristics of a WDM link in the WINMAN inventory.

- *Add EMS and delete EMS UCs.* These UC are intended to modify the network configuration (updating the WINMAN inventory) by adding or removing EMS information through an established communication link at the WINMAN southbound interface.

6.4.2.5 NM&R

The NM&R UCs support the WINMAN fault management functionality. This refers to the processing of alarm notifications and to the execution of actions triggered by such alarm processing. The WINMAN operator is the actor intended to acknowledge or clear alarms. Figure 6.11 depicts the NM&R UCs considered by WINMAN.

The NM&R UCs are:

- *Recover network UC.* This UC is used for rerouting all connections, bypassing faulty parts of the network.

- *Revert network UC.* This UC comes back to the original routing status, as it was before the recover network UC invocation.

- *Open alarm UC.* This UC handles alarms issued by an EMS, by an internal WINMAN application like the performance management, or by the topology manager when the network element status is changed.

- *Clear alarm UC.* This UC clears an alarm and removes it from the list of active alarms.

- *Provide alarm reports UC.* This UC creates reports based on selected attributes carried on the alarm notifications.

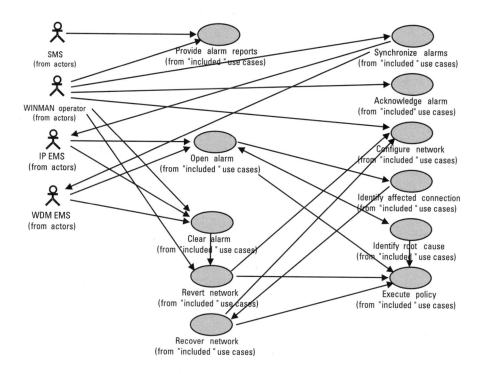

Figure 6.11 NM&R UCs.

- *Identify affected connections UC.* This UC finds all active connections affected by a given alarm.

- *Identify root cause UC.* This UC identifies the root or primary cause in a set of alarms.

- *Synchronize alarms UC.* This UC queries the EMSs for their open alarms for resynchronization of the active alarms in the WINMAN system. As a consequence of the execution of this UC, the alarms reflected in the WINMAN system will be the same as those existing in the corresponding EMS.

- *Acknowledge alarm UC.* This UC changes the status of an alarm to the acknowledged status.

6.4.2.6 Network Data Management

The network data management (NDM) UCs incorporate the functionality intended for performance management. These are the UCs providing means to collect and analyze performance data. The operator can request reports and configure the way WINMAN collects performance information. The configuration can be done on a counter-by-counter basis or an ICS basis. Policy-based management plays an important role. WINMAN intends to support a set of preconfigured data collection policies to be used by operators that do not want to have to understand the minute details. Figure 6.12 depicts the NDM UCs considered.

The NDM UCs are:

- *Process performance measurements UC.* This UC collects countervalues according to a defined policy. If the thresholds set on these counters are crossed, an alarm is triggered.

- *Provide PM report UC.* This UC produces performance management reports containing an analysis of the QoS supported by the corresponding ICS.

- *Calculate aggregations and statistics UC.* This UC calculates performance parameters as aggregation of the countervalues for different periods of time; this allows, for instance, the computation of the availability for a given ICS. Moreover, the UC calculates computed counters and statistics (compared to average and standard deviations) according to the customer's definitions. If some thresholds were crossed by the calculated values, alarms are generated and sent to the fault management UCs.

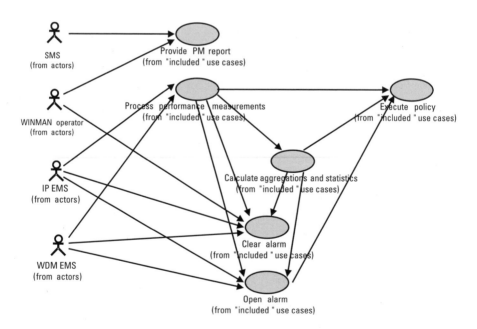

Figure 6.12 NDM UCs.

6.5 Summary

The WINMAN solution and hence the WINMAN system where it materializes is the result of a recursive process used in software engineering. It consists of five steps—the proposal of a business model, the requirements capture, the UCs specification, the drawing of the MSCs, and the design of class diagrams. Executing these five steps leads to the specification of the WINMAN system in terms of its main entities and the interfaces existing in between. The first three of these steps are in fact the subject of the chapter.

The business reference model identifies three roles—the customers, the VASPs, and the network service providers. These roles can be adopted indistinctly by any organization. Actors adopting these roles will be the entities dealing with the WINMAN solution and henceforth getting the competitive advantages that it provides.

Although the WINMAN solution is not intended to directly manage services like voice over IP or VPNs, it is paving the way to make them available by means of an ICS provisioning tailored to any user needs. This ICS is accomplished through an NP concept that allows the setup and maintenance of end-to-end IP paths over light paths using MPLS technology, supporting traffic and QoS parameters derived from SLAs.

Fault and performance management functions are also inherent advantages of the WINMAN solution. In the first case, the report and recovery of faults in the IP or in the optical path is accomplished. And for performance management, the monitoring and control of the QoS is the primary goal.

The WINMAN solution is therefore an integrated set of configuration, fault, and performance management facilities running on top of technology-dependent element-level management entities and providing this set of management functionality to both physical network operators or to service-level management processes through open interfaces.

The main functional requirements of the WINMAN solution are grouped according to the TOM of the TMF Forum. Specifically they cover NP, NIM, network maintenance, and restoration and NDM. Moreover, besides these standard requirements, others were also considered of special relevance, such as those concerning the data repositories and the system tests.

The system architecture in WINMAN is supported by the identification and description of appropriate UCs that illustrate the basic functionality of the targeted system. UCs are structured in a two-dimensional classification. One of these dimensions is the functional area covered, this is essentially the TMF-TOM classification augmented with one devoted to policies provision and another to task scheduling. The other dimension refers to the level of abstraction and can adopt the values of summary level, user level, or atomic level. In this way we have a grid of six rows and three columns containing more than 50 UCs.

The set of policies provision UCs specifies the functionality that WINMAN will possess in order to manage policies, such as the creation of new policies and the removal of old ones, and the system capacity to execute management task under policy control.

The set of scheduling UCs specifies the WINMAN solution facilities in terms of scheduling different types of tasks. A given task can be scheduled and at any time the scheduling can be modified or canceled.

The set of NP UCs expresses the WINMAN facilities in terms of managing the configuration of the network. The highest level entity representing the managed resource is here the ICS, an end-to-end transport facility with specific traffic and QoS parameters. The UCs state how these ICSs can be created, deleted, and modified. Also worthy of mentioning is that the UCs belonging to this set are specially suited for routing calculations.

The NIM UCs allow the management of both the physical and the logical resources of the network. Henceforth they show how to register different types of EMSs, as well as how to create, modify, or delete nodes and links of different technologies.

The NM&R set of UCs incorporate the WINMAN functionality for fault management. This consists of the recovery of connections under failure, which in turn is based on the capacity to identify faulty connections and the root cause of alarms.

The last set of UCs is constituted by the NDM UCs. These ones describe the functionality regarding the control and monitoring facilities that at the end will enable the calculation and the maintenance of the QoS.

References

[1]   Telemanagement Forum, TOM.

[2]   ITU-T Recommendation G.805, "Generic Functional Architecture of Transport Networks," March 2000.

[3]   ITU-T Recommendation G.852.2, "Enterprise Viewpoint Description of Transport Network Resource Model," March 1999.

[4]   TMF 509, "NML-EML Interface Business Agreement for Management of SONET/SDH Transport Networks," October 2000.

[5]   ITU-T Draft Recommendation M.3120, "CORBA Generic Network and NE Level Information Mode—Determination Version," October 2001.

[6]   ITU-T Contribution D.995 (WP 3/13), "G.cls Functional Model," March 2000.

7

The WINMAN System Architecture

7.1 Introduction

A system architecture design must fulfill several requirements. First of all, the architecture must be driven by the functional requirements of the system. The presentation of these requirements was the subject of Chapter 6. At the same time, the system architecture must exhibit a set of nonfunctional requirements that make it suitable for the system operation in real environments. Among these types of requirements we mention the most important ones as follows:

- *Openness.* This is an architecture with published APIs to the relevant actors, giving third parties the ability to implement their own management application. Wherever possible, interfaces specified by the relevant standardization fora (ITU, TeleManagement Forum, IETF, or OIF) should be adopted.
- *Flexibility.* The system architecture must be able to evolve with and accommodate new systems. In particular, seamless accommodation of other network management systems should be enabled.
- *Modularity.* The architecture must be able to have the capability to add or remove management blocks in a plug and play fashion without requiring major changes to existing components. This requires a clear separation of responsibilities between blocks.
- *Scalability.* The architecture should be able to cope with network evolution scenarios towards IP over WDM and growth of the network in terms of number of network elements and type of network elements.

- *Distributed.* The architecture must be able to operate on a platform in which distribution of the system is transparent to the constituting components.

- *Platform and persistency independence (essentially, database management system).*

Considering all of these statements, the WINMAN architecture is described at two levels. First of all, we identify a high-level architecture describing the role of each constituting system in a TMN domain [1] and mutual interactions. At this level, the systems are considered black boxes. At the second level, the internal architecture of the systems that comprise the WINMAN solution addresses such aspects as how these systems are built, which building blocks they have, what middleware is used, or how persistency is provided.

This chapter is structured around five sections. After this introduction, Section 7.2 is devoted to the high-level system architecture that is complemented by the domain model contained in Section 7.3. Also, an example is provided to illustrate the interaction between the components of this architecture. The internal architecture of the WINMAN system, the second description level mentioned earlier, is presented in Section 7.4. Nevertheless, this section is not entering in implementation details of the building blocks because this is the scope of Chapter 8. Section 7.5 presents the design aspects of the WINMAN GUI. Finally the chapter ends with the summary section.

7.2 High-Level Architecture

As presented in Figure 7.1, the high-level architecture of WINMAN comprises three different network management systems: the IP-NMS and the WDM-NMS at the technology dependent layer and the INMS above them. The user interface for the complete solution is also depicted in the same figure.

Only the simplest option supported by the solution—a simple system per management layer—is shown in Figure 7.1. Nevertheless, the architecture adopted in WINMAN can incorporate multiple SMSs over the northbound interface, multiple NMSs at the technology specific layer, and multiple EMSs under the southbound interfaces. Description of these three main subsystems follows hereafter.

7.2.1 The Interdomain Network Management System

The interdomain network management system (INMS) is devoted to carrying out the integration of the management functions of the underlying technology-specific management systems. Each of these technology-specific management

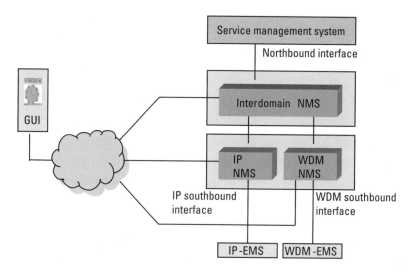

Figure 7.1 WINMAN high-level architecture.

systems is seen as the header of a technological domain. This is why we call the entity on top of them an interdomain manager. The INMS is technology agnostic in that its representation of the network is not tied to any specific technology. Through a properly abstracted information model, the INMS is able to use the domain-specific NMSs for the provisioning and monitoring of end-to-end connections without any knowledge about network details except for its layered structure. This means that the INMS does not make any assumptions about the type of network managed by the domain-dependent NMSs; it only requires them to support a common interface for the establishment and releasing of connections and for the notification of fault- or performance-related events. This common interface is devised for connection-oriented networks following the main focus of WINMAN.

In this book we only cover the configuration management functionality of WINMAN. This functionality enables single point access to provisioning tasks and to end-to-end views of connections and their underlying infrastructure and facilities, independent of the technology domain. Specifically, the main functions that the INMS performs in the area of configuration management are:

- Serving the requests from the SMSs for the provisioning, modification, and deletion of the ICS, which, as defined in the previous chapter, consist of IP over WDM connectivity with QoS guaranties;

- Configuration of the end-to-end ICSs across the different technology domains, designing the end-to-end route as an ordered set of subnetwork connections;

- Presentation of an end-to-end view of the services to the operator showing the logical hierarchy of all network resources (nodes and links) constituting the end-to-end connection;

- Maintenance of an inventory of all the network resources relevant at this layer with their status and their hierarchical relationship— this system only has knowledge about subnetworks at each of the domains and about points of interconnection at the edge of those subnetworks;

- Coordination of the restoration process when services become faulty or fall under acceptable QoS levels;

- Provision of updates to the fault and performance management functions, reflecting the changes in the network configuration and the establishment of new services.

7.2.2   The WDM-NMS

The WDM-NMS performs the management of the WDM transport network layer and offers a technology-independent management view of the WDM network domain. It relies on one or several WDM-EMSs that provide a first-level abstraction of the network and deal with vendor-specific aspects of the network elements. The WDM-NMS incorporates full configuration management functionality so that it can be deployed as a stand-alone management system capable of managing a WDM network by its own.

The configuration management functionality supported by the WDM-NMS is the following:

- Provisioning of optical paths, including their respective protection paths;

- Wavelength routing, including the capability of wavelength transition and implementation of the designed optical path in the network elements;

- Maintenance of an inventory containing the WDM resources and WDM connections;

- Coordination of the actions needed for the recovery of paths at the optical network;

- Discovery of network resources in terms of network elements, ports per network element, fiber connections between nodes, wavelengths per fiber, transport capacity per wavelength, and protection mechanisms. The notifications of change of any resource are forwarded to the INMS only if the modification is affecting the view of the network at this higher level.

7.2.3   The IP NMS

The IP policy-based NMS carries out the management of the IP network layer and offers a technology-independent management view of the IP network domain, adapted to the connection-oriented approach on which WINMAN relies. This adaptation has been enabled by the adoption of MPLS [2] as the basic networking technology managed by WINMAN. As pointed out in Chapter 3, MPLS enables the control of the path taken by the IP packet flows and provides the grounds for effective traffic engineering. In WINMAN, each IP connectivity service is mapped and transported over an LSP.

The IP-NMS relies on one or several IP-EMSs for the management of those aspects of the network elements not relevant at the NML. As in the WDM case, the IP-NMS is able to work as a stand-alone system for the end-to-end management of the IP network.

The configuration management functionality for the IP-NMS is the following:

- Design of connections inside its own subnetworks;
- Establishment and maintenance of end-to-end QoS connectivity services, including the setting up of MPLS LSPs;
- Maintenance of an inventory with the IP resources and MPLS connections;
- Coordination of the network recovery process when it affects only the IP network;
- Discovery of network topology. Like in the WDM-NMS, the notifications of resource changes received from IP-EMSs are transmitted to the INMS level only if the modification can affect the view of the network at this higher level.

7.2.4   The GUI

WINMAN has only one GUI that interacts with the three previously described systems. It supports the access to information and functions of the INMS, the WDM-NMS, and the IP-NMS.

The GUI is conceived as a thin Java client so that it is possible to use it through the Web. The GUI embeds a middle tier, the *view manager*, which works as a gateway between the graphical client and other internal application modules of WINMAN.

7.2.5   Component Interaction Example: A Connectivity Service Provisioning Scenario

The WINMAN architecture can be better understood by looking at a provisioning scenario that illustrates the functionality of each system and how the systems interact with each other.

This scenario describes the process of provisioning an ICS, which is the basic service provided by the WINMAN system.

Figure 7.2 shows a detailed view of the scenario that is triggered when the operator uses the WINMAN GUI to request the creation, implementation, and activation of a new ICS. Summarizing, the INMS finds out that there are not enough resources in the current IP network and issues a request to the WDM-NMS to provide one single optical connection. Once this has been done, the INMS issues a request to the IP-NMS to complete the provisioning of the new ICS.

As seen in Figure 7.2, the EMSs are also involved in the process. They are the gateways between WINMAN and the network elements, where all necessary connections are actually set up.

What is remarkable in this scenario is the relationship between the IP and WDM networks. In this case, the lack of bandwidth in the IP network is solved with the establishment of a new optical path that connects two IP routers. This results in an IP topology change that is reported to the IP-NMS in step 5 of the scenario. The additional available capacity allows WINMAN to find an

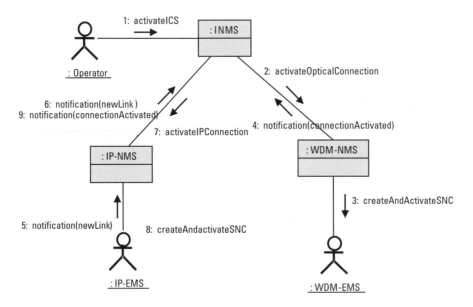

Figure 7.2 ICS provisioning scenario.

appropriate path to satisfy the quality requirements of the ICS and to complete the provisioning process by activating the ICS.

7.3　Domain Model

The domain model describes the entities that participate in the management processes, giving the specific views existing at each domain interface.

WINMAN is focusing on a connection-oriented IP domain model, according to the MPLS technology. Based on the G.805-layered architecture [3], we identify two layers: IP-MPLS and WDM. The coexistence of this model with the conventional IP connectionless technology is out of the scope of this book but we would have had three layers if it would have been considered.

Moreover, we suppose that the IP-MPLS domain starts at the customer edge IP equipment and thus when traffic enters the WINMAN responsibility (provider edge IP equipment), it is already marked as MPLS.

7.3.1　Common Entities

The model is characterized by the set of entities described hereafter and represented in Figure 7.3. These entities can later be mapped to each specific domain, as described in Section 7.3.2.

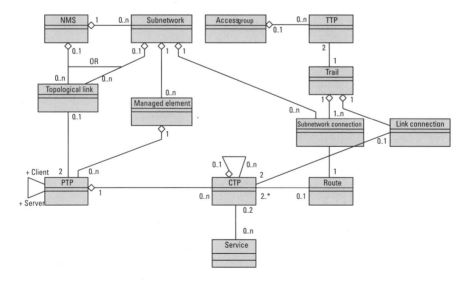

Figure 7.3　Generic domain model adopted in WINMAN.

- *CS.* The CS is the basic service delivered by the IP-MPLS or WDM network and managed by WINMAN. The service is delivered either by the IP-MPLS network meeting some QoS goals, resembling a connection-oriented circuit, or by the WDM network.

- *EMS.* The EMS represents the abstraction of the subnetworks managed by the technology-dependent management system.

- *Termination point (TP).* A TP is a logical abstraction of an end point (actual or potential) of a topological (physical) link or a subnetwork connection. The TP is contained within a managed element (ME), an entity that is described later.

- *Physical termination point (PTP).* A termination point that is an actual or potential end point of a topological (physical) link shall be abstracted as a PTP. Essentially, this is a representation of a physical port.

- *Connection termination point (CTP).* A CTP is an actual or potential end point of a subnetwork connection.

- *Topological link.* A topological link is a physical link between two PTPs. A topological link has a name and references to the two PTPs. A topological link reported by an EMS to a NMS will be between two MEs managed by the same EMS. Depending on the capabilities of the EMS and the MEs, the autodiscovery of topological links may or may not be allowed by the EMS.

- *Link connection.* A link connection represents the transparent capacity of transfer information characterized by a given signal identification between two fixed points.

- *SNC.* An SNC relates CTPs. An SNC provides a transparent end-to-end connection through or within a subnetwork. The SNC may be created, deleted, or modified by the INMS or by domain-specific NMS and is implemented by means of one or several EMSs. The SNC is contained in the subnetwork.

- *Route.* The route of a subnetwork connection shall be represented as an ordered series of CTP names through which the subnetwork connection traverses, including the working and the protection route. The protection route is optional.

- *ME.* An ME is an abstraction used to represent network elements visible across the interfaces.

7.3.2   Mapping Between the Model Entities and the Managed Resources

The WINMAN domain model is generic, thus providing an abstraction of the physical elements of the IP over WDM network by capturing in a uniform representation the relevant attributes, for management purposes, of any network entity. While this abstraction enables technology-agnostic processing, it is based on a mapping that is specific for each of the three management levels considered in WINMAN. Table 7.1 summarizes this mapping.

Table 7.1
Mapping of the Model Entities to Physical Resources

Entity	IP-NMS	WDM-NMS	INMS
CS	MPLS LSP between two end points of an IP subnetwork	Optical channel path between two end points of a WDM subnetwork	MPLS LSP between two customer-facing end points; also referred as an ICS
EMS	IP-EMS and the IP subnetwork it manages	WDM-EMS and the WDM subnetwork it manages	Not present
TP	PTPs and CTPs represent different resources		
PTP	Router interface	OXC or OADM port	Router interfaces and OXC/OADM ports
CTP	FEC assigned to a LSP	Wavelength at a OXC or OADM port	Not used
Topological link	IP point-to-point link between two adjacent routers	Fiber span between two adjacent OXCs/OADMs	Fiber span between a router and a OXC/OADM (interdomain link)
Link connection	Not present	Single channel of a WDM signal in a topological link	Not present
SNC	An LSP segment or an end-to-end LSP	Path between any two WDM CTPs	LSP between two end points of an IP network, or optical channel between two end points of a WDM network
Route	Sequence of IP router interfaces	Sequence of port/wavelength pairs	Sequence of IP router interfaces
ME	IP router	OXC or OADM	IP router (WDM network elements are not visible at this layer)

7.4   Generic Management System Architecture

This WINMAN architecture looks at the NMSs as derived from a generic solution handling the functional management areas: fault, configuration, and performance management. This generic NMS handles in a very similar way any type of domain whether it is WDM, IP, or INMS. Therefore, this architecture is also extensible to manage any other type of technology domain (e.g., ATM or SDH).

Figure 7.4 shows the system class diagram for the three types of NMSs, which are inherited, from the abstract or generic NMS. All of the NMS functionality is contained in the abstract NMS. This functionality is described by the following subclasses:

- GUI;
- Fault management, depicted as FM in Figure 7.4;
- Configuration management, depicted as CM in Figure 7.4;
- Performance management, depicted as PM in Figure 7.4;
- Database, depicted as DB in Figure 7.4;
- Northbound, depicted as NB Figure 7.4;
- Southbound, depicted as SB Figure 7.4.

Each of the specific NMSs contains all of the functionality of the components and therefore provides a full and independent management system. Under this point of view, WINMAN adopts the polymorphism principle so that each time a new type of NMS is inherited for a new type of network, the same models could be used without changing the architecture itself.

This view also shows the difference between the networks and the software management components. The different types of NMSs are related by inheritance to the abstract NMS object. The different components of the network management functionality are associated to the abstract NMS by means of containment relationships.

7.4.1   The Three-Tiers Concept

The three-tiers concept [4] has been developed in the world of browser applications when the need for independence between the persistence layer and server process to the business logic and the presentation is a must. The same need for independence between layers was identified for the WINMAN generic architecture.

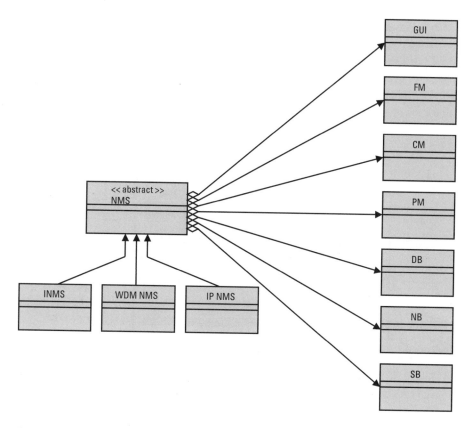

Figure 7.4 Class diagram for the generic NMS architecture.

When looking at the WINMAN NMS architecture at the horizontal view, we can see the three-tiers concept made reality. In fact, as shown in Figure 7.5, WINMAN is decomposed horizontally in the client/presentation layer, the middle tier/business logic layer, and finally the server/persistency layer. A description of each one follows.

7.4.1.1 The Client/Presentation Side

This layer contains the GUI itself and the view manager. These two components permit there to be a thin client that enters the GUI logic in the view manager. The GUI can be independent from the server, thus having one point of contact to all of the WINMAN functionality—the view manager.

7.4.1.2 The Middle Tier/Business Logic Side

This layer contains the business logic such as the NMS processes, data manipulation, rule-base systems, and the communication with the EMSs for the

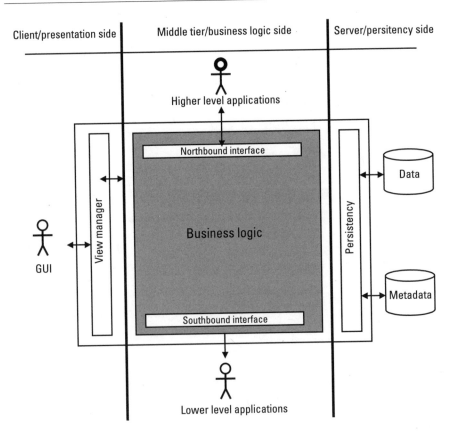

Figure 7.5 The three-tier architecture of the NMS.

provisioning processes. This is done in order to enable different rules and algorithms to be plugged into the system.

7.4.1.3 Server/Persistency Side

This layer contains the persistency component and the database itself in order to keep the system database independent and to allow different persistence implementations such as relational databases, object databases, or file systems.

In Figure 7.5, we can also observe that the WINMAN NMS communicates with its environment through four types of interfaces, as described hereafter.

7.4.2 The Northbound Interface

This component is a generic interface to the northbound customer applications. This interface enables the WINMAN NMS to stay independent of the type of customer application that wants to use and interact with it. This component gets the customer's commands and requests using a well-known protocol and translates it to the actions inside the WINMAN NMS. From the other end, this interface receives all of the events from the WINMAN NMS and sends them northbound to the customer, using the same communication protocol. The customer of the northbound interface can be an external application like the SMS layer or another higher level NMS like the INMS, with respect to the WDM-NMS or IP-NMS.

7.4.3 The Southbound Interface

This component is a generic interface to the NMSs. This interface enables the WINMAN NMS to stay independent of the type of network that it manages. This component gets the events from the EMSs or lower level NMSs using a well-known protocol and presents the events to the WINMAN components. From the other end, this interface gets queries and requests from the WINMAN modules and using the same communication protocol sends them to the EMSs or lower level NMSs.

7.4.4 The View Manager Interface

The view manager is a component that links the WINMAN processes to the GUI, in order to keep the GUI as independent as possible from the process in the server/middle tier. This component will get all of the actions initiated by the WINMAN operator in the GUI and will forward them to the relevant component in WINMAN. On the other hand, all of the answers from the WINMAN components are translated by the view manager to a simple structure known by the GUI so that it can process and display the data without knowing the specific structure inside each of the WINMAN components.

7.4.5 The Persistence Layer Interface

This interface is responsible of the persistency of the WINMAN data. It gets all of the data processed by the NMS components and stores it. At the same time, this component restores all of the relevant data (and, if needed, all of the relevant metadata) from the storages and brings it to the different components. This component enables the WINMAN solution to stay independent from the storage type, which can be any database or any file system (local or remote).

7.5 The GUI

The WINMAN-integrated GUI acts as a general-purpose tool, which can be used for:

- Conveying the management functions to the user/operator in an abstract and easily perceptible way;

- Testing WINMAN external interfaces;

- Testing and verifying the overall system functionality.

A basic requirement for the WINMAN GUI is that it concerns a thin client that can test and control all WINMAN functionality. The WINMAN GUI is a Java-based, platform-independent application for operators of the underlying management systems. It provides a common look and feel for managing the underlying subsystems, whose technologies, interfaces, and protocols may be very different from each other. The idea behind the WINMAN GUI is to split the management tasks into small functional components, which will together cover all of the required functionality, with a common look and feel as well as a common starting point for the system administrator.

In the WINMAN's three-tier functional architecture, the GUI assumes the *thin* client role. A thin client holds minimal functionality. However, the client runs mainly presentation and communications functions, which allow one to access services over the network and display the results to the user. The majority of data management and business logic intelligence is executed outside the GUI, in the server layers, as discussed in previous sections.

7.5.1 Model-View-Controller Architecture

The model-view-controller (MVC) [5] pattern plays a fundamental role in deploying GUI objects. It is closely interwoven with object-oriented design (OOD). The three core abstractions of MVC are:

1. *The model.* This holds all data relevant to an entity or process and performs data processing. The data is independent of the component's visual representation. The model notifies the view that its state has changed, and the view needs to be redrawn.

2. *The view.* This accesses and displays data contained in the model. It knows about the model only to the degree allowed by the model's *public property interface.*

3. *The controller.* This is the glue between view and model. It reacts to gestures and events in the view enacted by the user that may result in the update of data in the model. The controller knows about the model only to the degree allowed by the model's *public behavior interface.*

The MVC pattern can be contemplated in combination with the previously mentioned three-tier architecture. The view is on the top layer, and the model is located at the two lower layers.

7.5.2 GUI Functional Components

The GUI subsumes four functional views, which deal with different aspects of network monitoring. These are depicted coarsely in Figure 7.6 and mentioned in brief in the following paragraphs.

7.5.2.1 Configuration Management Functional Component

The configuration management component is essentially made by the network map visualization component. The network map can adopt either a *tree view* or a *mesh view*, depending on the topology of the network structure.

- The tree view shows a hierarchical picture of subnetworks and network elements, which might correspond to a physical or logical network hierarchy.
- The mesh view solves the problem generated from the fact that the area of a computer screen cannot visibly represent large networks including thousands of nodes. The mesh view is designed so as to allow the user or operator to select physical or logical entities (i.e., subnetworks and components of each subnetwork and even network nodes that he or she wants to query for network information). It allows

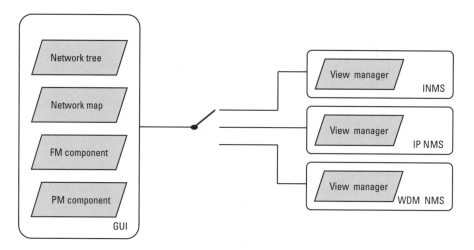

Figure 7.6 The WINMAN GUI high-level architecture.

selection capabilities based on filtering of network elements using specific assignable attributes.

7.5.2.2   Fault Management Functional Component

The fault management view displays the alarms and notifications that are sent by the underlying management system. The user manipulates the received alarms by acknowledging, assessing, or clearing them. The alarms are displayed in tabular form and have various color markings for severity indication.

7.5.2.3   Performance Management Functional Component

This component is responsible for visualizing performance-related information concerning certain parameters of physical or logical entities. This information is recorded in the corresponding modules of the system. Data is displayed in charts where the time interval is either arbitrary (user defined) or predefined. In the latter case, the charts are of the form of system performance statistical measurements.

7.6   Summary

The high-level architecture of WINMAN consists of three systems—the INMS, the IP-NMS, and the WDM-NMS, which communicate between each other and with external systems via open standard interfaces. The INMS is devoted to carrying out the integration of the management functions of the underlying technology-specific management systems. The WDM-NMS performs the management of the WDM transport network layer and offers a technology-independent management view of the WDM network domain. The IP-NMS carries out the management of the IP network layer and offers a technology-independent management view of the IP network domain, adapted to the connection-oriented approach on which WINMAN relies. Moreover, only one GUI interacts with the three previously described systems supporting the access to information and functions of the INMS, the WDM-NMS, and the IP-NMS.

For the internal structure of the systems, a generic architecture based on common building blocks for the three WINMAN systems was defined to speed up implementation and facilitate future extensions to other technologies. This common NMS architecture defines a number of components devoted to specific functions in the configuration, fault, and performance management areas. This generic approach requires in practice a flexible, extensible, and powerful domain model that should be able to represent any entity existing in the network. The selection of technology-neutral concepts and standards is the basis for this domain model. Also characterizing the architecture of the generic WINMAN

NMS, we can see that it is a three-tear conception; that is, WINMAN decomposes into the client/presentation layer, the middle tier/business logic layer, and the server/persistency layer.

From point of view of communication with the external environment, the generic WINMAN NMS has four types of interfaces—the view manager interface, the persistency layer interface, the northbound interface, and the southbound interface. The view manager interface links the NMS WINMAN to the GUI and keeps the GUI as independent as possible from the process in the server/middle tier. The persistency layer interface enables the WINMAN solution to stay independent from the data storage type, which can be any database or any file system. The northbound interface is offered to clients like the SMS layer or another higher level NMS, such as the INMS. The southbound interface is based on services offered by external EMSs or lower level WINMAN NMSs.

References

[1]   ITU-T Recommendation M.3010, "Principles for a Telecommunications Management Network," February 2000.

[2]   Rosen, E., A. Viswanathan, and R. Callon, "Multiprotocol Label Switching Architecture," *IETF Request For Comments 3031,* January 2001.

[3]   ITU-T Recommendation G.805, "Generic Functional Architecture of Transport Networks," March 2000.

[4]   Edwards J., *3-Tier Server/Client at Work,* Revised Edition, John Wiley & Son Ltd, 1999.

[5]   Buschmann, F., et al., *Pattern-Oriented Software Architecture: A System of Patterns,* John Wiley & Son Ltd, 1996.

8

Management System Design and Implementation Issues

8.1 Introduction

Following the generic description of the WINMAN system provided in Chapter 7, we will now describe the details of the components that form the aforementioned system. In short, we concluded that the WINMAN system can be decomposed into three subsystems—the INMS, the IP-NMS, and the WDM-NMS. Also, we stated that besides their different technological scope, the conception of the three systems derives from a common generic NMS system. Regarding the generic NMS, we also stated that it is based on a three-tier concept, and it exhibits four types of interfaces with the external environment. Up to this point we have not mentioned anything else about its internal structure. This is the starting point of the present chapter, the objective of which is to point out the design and implementation of the three types of NMSs. This description is based on the proposal outlined in [1] that identifies the architecture of the previously mentioned generic NMS and states that it is constituted by the system components depicted in Figure 8.1.

It is worth noting that the components of Figure 8.1 are strictly necessary to provide the generic NMS its full management capacity in the configuration management area. Therefore, we want to emphasize that neither in this chapter nor in other chapters of the book are we going to detail how we have implemented the NMSs exhibiting the fault management and the performance management functionality. We have decided to use the pages of this book to go more in depth on the system conception rather than on the whole system implementation details.

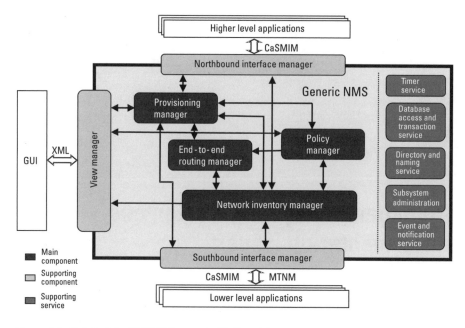

Figure 8.1 Detail of the WINMAN generic NMS architecture.

Each component will be individually described. Within each description we also present a UML diagram depicting the environment in which each component is operating, the interfaces that it supports towards the other components, and, if needed, the interfaces that are being used from other components. Finally, where the description of a given component of the generic NMS is complete and major enhancements or adaptations of the generic component are needed to fulfill the specific requirements of the INMS, the IP-NMS, or the WDM-NMS, these adaptations are described in separate subsections.

The chapter is structured in 10 sections. Besides this introduction and the summary ending the chapter, a section is devoted to each of the eight NMS system components.

8.2 Provisioning Manager

The process of provisioning new services or connections in communication networks tends to be complex and error prone. The large number and the variety of network elements involved, most often with different and sometimes conflicting capabilities, demands that special care be taken in all of the actions carried out to fulfill a provisioning request. The efficiency of this process can be improved if a single entity takes care of the coordination of all necessary activities. In WINMAN, the provisioning manager assumes this role.

The main function of the provisioning manager is hence to control the provisioning of connectivity services in the WINMAN-managed IP over WDM network. This component receives the requests issued from the WINMAN operator through the GUI or from the SMS application through the northbound interface and triggers, via the southbound interface, the required configuration actions on lower level applications to activate the services. According to Chapter 6, there can be three types of connectivity requests: *provide ICS/CS, modify ICS/CS,* and *delete ICS/CS.*

These requests are passed to the provisioning manager using the proper operations defined in the TMF connection and service management information model (CaSMIM) interface [2], designed as a technology-agnostic interface suitable for its use between SMSs and network management systems, and between interdomain management systems and technology-specific management systems.

As shown in Figure 8.2, separate interfaces have been defined for northbound requests from the SMS (*INorthboundInterface*) and for GUI requests from the WINMAN operator (*IViewManager*) to better cope with the specific characteristics of each type of request. However, the same information model has been adopted and the same set of functions is supported.

Providing a connection means selecting the resources that will be allocated to that connection and configuring those resources in the network to physically

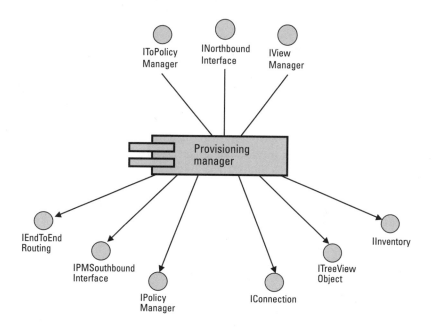

Figure 8.2 The Provisioning manager context.

establish the connection. As a first step, when the connection entity is created in the system, its route is computed but no actions are performed in the network until the implementation stage. In case of a scarcity of resources, it may happen that a suitable route for satisfying the quality requirements of the connection request cannot be found. The provisioning manager of the INMS reacts to this situation by requesting the activation of new capacity. In an IP over WDM network, capacity is obtained by setting up a new optical path between two IP routers. The provisioning manager will create as many additional optical paths as are needed to satisfy the request quality constraints.

While in the creation phase of a connection, the resources are logically reserved, though only in the WINMAN databases. In the implementation stage, all network elements are actually configured to support the service; however, it still cannot carry traffic.

The final step of the provisioning process is the activation of the service, after which it becomes fully operational. The provisioning manager supports both step-by-step provisioning and single-request provisioning. It must be noted that typically all steps (creation, implementation, and activation) are done automatically after a single SMS/operator request.

The deletion of a service requires traversing the same states in the opposite direction. The deactivation of the connection causes the cancellation of the services delivered over such a connection. Resources are then released both in the network elements and in the NMSs. Finally, all information about the connection is deleted.

Modification of a service is done as a sequence of delete and provide operations and always causes service interruption. A more flexible approach, based on the survivability features that IP routers provide, is being considered.

It should be noted that in the network provisioning processes of an ICS, more than one provisioning manager instance is involved. In fact, each of the INMS, IP-NMS, and WDM-NMS contains its own instance of the provisioning manager, derived with few modifications from the generic provisioning component. These different instances interact with each other in the described processes. For example, when the INMS provisioning manager decides that a new optical connection is required it delegates on the WDM-NMS provisioning manager, which then coordinates the activation of such a connection.

The provisioning manager relies on the end-to-end routing component to support advanced features associated with a service. The provisioning manager receives the required features and transparently forwards them to the routing component, which is then responsible for considering those features while computing the route for the service. In this way, the system can be enhanced with new functionality without affecting the provisioning component.

Authorization-support capabilities have been implemented through the policy manager validating all requests before proceeding to honor them. This checking is done at the very beginning of the provisioning process to prevent further processing of requests that are unacceptable according to the user-defined policies or which are technically inconsistent.

Finally, this component is able to receive notifications from lower level applications and to send them upwards to higher level applications, in order to keep synchronized the information about services and their states in all systems within the network management hierarchy. The CORBA notification service is used for this purpose.

8.3 End-to-End Routing Manager

This component is in charge of calculating the route between the two end points of the connection [3]. When a provisioning request is received in the system, the end-to-end routing manager makes use of the network graph representing the current network topology to quickly find the most efficient route between two given end points. The necessary network topology information is retrieved from the network inventory manager. Route calculations are only requested by the provisioning manager component.

The end-to-end routing manager can base its calculations on *ad hoc* policies. These policies, for example, can exclude certain links from routing calculations or give preference to others to find more efficient routes in terms of available bandwidth, hop count, economical cost, or even a combination of these factors. Figure 8.3 depicts the end-to-end routing manager context.

Table 8.1 presents the main characteristics exhibited by the routing algorithm, explicitly showing those that are generic and those that are system specific.

8.4 NIM

The NIM or inventory component is designed to have a full view of the network. Its responsibility is to know the different elements of the network, their roles, and their connections to other elements. The inventory component should have a full view of all of the subnetworks that the configuration module is managing. The inventory component roles are:

- Maintain an updated and persistent view of the network;
- Navigate on the network associations between networks and network elements;
- Display the network tree view.

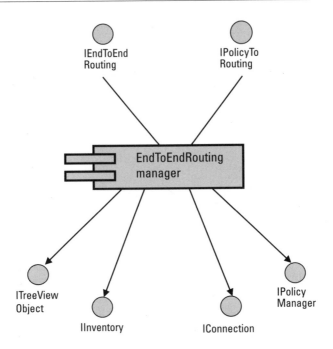

Figure 8.3 The end-to-end routing manager context.

Table 8.1
Characteristics of the End-to-End Routing Algorithm

Generic	INMS Specific	IP-NMS Specific	WDM-NMS Specific
On-line service algorithm	Ability to find WDM capacity to fulfill IP connectivity requests		First-fit wavelength selection
Shortest paths first selection			No wavelength conversion considered
The same path for both directions whenever possible			Physical impairments impact minimization
Path protection calculation enabled			Nonblocking OXCs and OADMs
Admission of bandwidth and QoS constraints			The same primary and backup wavelength
			Fiber diversity allowed

8.4.1 The Inventory Model

The inventory has two views: a physical one and a logical one. In technology-dependent NMSs (IP-NMS and WDM-NMS), both physical and logical presentations exist side by side. The physical presentation includes the subnetwork, the managed elements (like routers or OADMs), and equipment such as cards, chassis, and PTPs. The logical view presentation includes the subnetwork connections, services, and CTPs. This is an extendable model that can be augmented to support any other type of network technology.

The INMS inventory only has connections and services. However, in order to have a full view of the connection, the entities that create the connection, which are end points and links, are mirrored from the IP-NMS and WDN-NMS inventory components in the INMS inventory.

8.4.2 Querying the Inventory

The inventory supports three interfaces towards other components, as shown in Figure 8.4. Each is designed to support different types of queries:

- IInventory—general information about the objects and their attributes;
- IConnection—information about the connection;

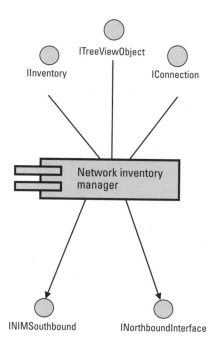

Figure 8.4 The network inventory manager context.

- ITreeViewObject—information about the parent-child hierarchy of the objects.

In the technology-dependent NMSs, the inventory gets all of the data from their relational databases, which have the same table structure regardless of the domain type (IP or WDM). The identity of the querying entity (GUI, INMS, or other components) is transparent to the inventory and thus keeps the component as an independent entity.

In the INMS inventory, there are two types of queries: one that relates to the INMS objects and another that relates to the underlying technology objects. In the first case, when the entity belongs to the INMS, the inventory behaves in the same way as in the IP-NMS or WDM-NMS inventory and receives all the required information from the relational database. In the second case, when the object belongs to the IP-NMS or WDM-NMS, the inventory gets the information from the corresponding inventory components using the INMS southbound interface.

As with the technology-dependent inventories, the identity of the querying entity (GUI, INMS, or other components) is transparent to the INMS inventory and thus keeps the component as an independent entity.

8.5   Policy Manager

The policy manager role is twofold. First, it is the WINMAN component responsible for the execution of actions derived from policies to be enforced in the managed network. Second, it is responsible for the management of the life cycle of these policies.

The policy execution can in turn consist of two types of processes. The first type consists of the validation of provisioning requests received by any of the three instances of the provisioning manager [4]. For this purpose, the policy manager analyzes the CaSMIM connection parameters carried in the provisioning requests, and applies validation policies previously established by the WINMAN operator. Through validation policies, the WINMAN operator can manage the resources when convenient. The validation constraints that the WINMAN operator can apply through validation policies are based on connection parameters and basic timing criteria. The second type of execution process consists of applying changes on the routing graphs managed by the end-to-end routing manager, according to routing policies that are also specified by the WINMAN operator.

To carry out the management of the life cycle of the policies, the policy manager offers a policy edition tool in order to add, modify, or delete policies for any of the three WINMAN NMSs. The policy edition process is invoked through the WINMAN GUI. Figure 8.5 depicts the policy manager context.

The policy manager is a unique component instance that validates provisioning requests and applies routing policies either for the INMS, the IP-NSM, or the WDM-NMS. Therefore, specific system adaptations are not needed, but specific policies for WDM-NMS, IP-NMS, and INMS levels that will be properly discriminated are necessary.

The policy manager is based in the Ponder policy specification language and development framework [5]. The WINMAN operator specifies policies in the high-level Ponder syntax that are then translated in Java classes and stored in a LDAP repository. These policies are extracted and evaluated when necessary by the policy decision point, another subcomponent of the policy manager, developed in WINMAN for this purpose. The Ponder authorization policies are used to represent our authorization policies, and the Ponder obligation policies are used to express our routing policies. The WINMAN operator can transparently manage these policies through the WINMAN GUI, bypassing the Ponder GUI.

8.6 View Manager

The view manager is a gateway component that links the GUI to the WINMAN server. Its main role is to enable the GUI to operate independently and to view

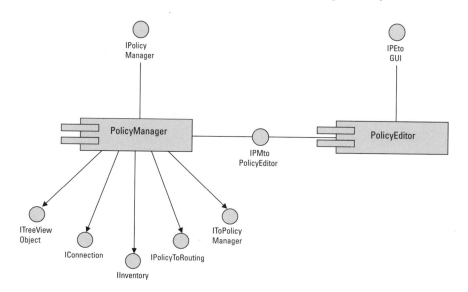

Figure 8.5 The policy manager context.

the entire WINMAN server as a single component. In this way, the GUI has one point of contact to the WINMAN server, regardless of the required functionality or the viewed domain.

As shown in Figure 8.6, the view manager has a published interface with generic operations that allow the GUI to retrieve data from the server, update the data in the server when the operator entered data manually, and start a provisioning process by creating an object.

These operations allow the GUI to interface with the inventory, the provisioning manager and the policy manager without knowing in which specific system they are.

Another important role of the view manager is to unify the object structure. The GUI uses a flat object that has only one level of hierarchy, while the WINMAN server uses nested objects. The view manager gets the WINMAN objects (flat ones) from the GUI, recreates the nested structure, and sends them to the WINMAN server components, and vice versa. The view manager gets the nested objects from the inventory and creates the flat WINMAN objects for the GUI.

In order to keep these two mapping mechanisms (mapping to the server components and mapping the server-nested objects) and allow them to be

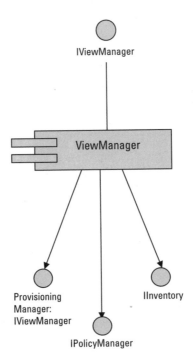

Figure 8.6 The view manager interfaces.

augmented without any code changes, the view manager uses extended markup language (XML) files to describe them. There are two types of files:

1. *Input files.* Mapping the GUI request to the server-related component. It maps the GUI flat object to the server-nested objects.

2. *Output files.* Mapping the server response to the GUI. It maps the server-nested object to the GUI flat object.

In order to allow the GUI to connect to the IP-NMS and to the WDM-NMS, there is a view manager component for each domain. However, these components are identical. There are no changes of behavior between the IP view manager and the WDM view manager. In the same way, the view manager component is an independent component and can be connected to any other domain type.

At the INMS level, the view manager has another role; if the request from the GUI needs to be forwarded to the technology-dependent inventories, it communicates directly to the INMS southbound interface for retrieving the data. This behavior is also defined in the XML mapping files, and therefore the INMS view manager component can be the same as in the IP and WDM systems.

8.7 Northbound Interface Manager

A focus point of the WINMAN architecture is to provide NMSs in multivendor and multitechnology environments. This goal is attained by having multiple instances of NMS systems, one for every different domain managed by the WINMAN solution. In order to provide access to network services in a technology-agnostic style, the functionality of each domain-specific NMS is accessed through the INMS system. In order to have a simple implementation of such a design, the communication between each of the various cross-domain managers (IP-NMS and WDM-NMS for the current implementation) must be based on a generic interface that provides isolation from the specifics of each technology. This interface is implemented by the northbound interface manager (NIM) component.

The primary role of the northbound interface is to provide a single entry point to clients that wish to gain access to the services provided by the NMS in a way that is independent of the underlying technologies. These requirements are in line with concept of the TMF CaSMIM standard [2].

The CaSMIM is a set of CORBA interfaces that enables the user to provide connection services across diverse technologies. The implementation of the NIM component exports a CORBA set of interfaces that implement the

CaSMIM standard. The incoming CaSMIM calls are then translated by the interface component into technology-specific method invocations that are compatible with the current NMS system.

The interfaces of CaSMIM are organized in a set of profiles. The methods are grouped in these sets according to the functionality they provide. The profiles currently defined are the following:

- *Basic.* The basic profile must be supported by all conforming implementations. It specifies the basic functionality that is needed for providing connectivity and connection handling.

- *Performance.* The performance profile is optionally available only in conjunction with the fine grain profile, but clients may gain significant performance improvements when using it.

- *Monitor.* The monitor profile is required for clients who desire real-time monitoring of the health or status of connections.

- *Topology.* The topology profile is provided by implementations that expose information about their supporting subnetworks or network element implementations.

- *Customer.* The customer profile includes support for handling customer-related information on connections. This can be useful for establishing links to customer service OSSs. However, the information is not used for actual connection management.

- *Schedule.* The schedule profile includes information and operations for implementations offering scheduling and historical records.

- *Fine grain.* The fine grain profile represents connections, links, access groups, and terminations as CORBA objects, in addition to their representation as second-class objects (coarse grain model).

- *Completion.* The completion profile gathers functions that are typically not included in other profiles. To claim conformance to the completion profile, all other profiles except the fine grain profile have to be supported.

For the current implementation of the NIM, a subset of the aforementioned profiles was used: the basic, customer, monitor, and topology. Moreover, the implementation contains methods from the classes listed in Table 8.2.

All of the interfaces listed in Table 8.2 are subclassed by Casmim::Subnetwork. This means that the northbound manager, as it is depicted in Figure 8.7, implements part of the methods defined in this interface.

Table 8.2
CaSMIM Interfaces Used in the NIM

Casmim::AccessGroup2	Casmim::AdministeredObject2	Casmim::Connection2
Casmim::ManagedObject2	Casmim::Subnetwork2	Casmim::Termination2
Casmim::Link2		

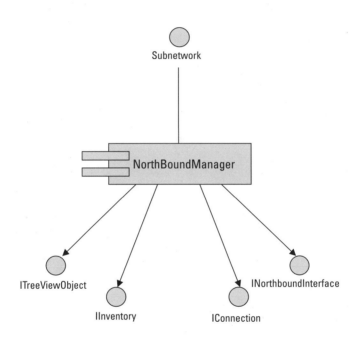

Figure 8.7 The northbound manager context.

The current implementation also supports receiving and sending of notifications through the *Common::Notification::Generator* and the *Common::Notification::GeneratorFactory* interfaces.

According to the standard [2], there can be two ways of implementing the CaSMIM interface—the fine grain and the coarse grain approach. In the first, each network element in the system is represented by a discrete CORBA object. While this approach is straightforward in its implementation, it results in a large performance overhead because all of the representations of network elements must be mapped in the memory regardless of whether they are currently used or not. In the coarse grain approach, currently used in WINMAN, each object is represented by a unique CaSMIM identifier. While this approach provides superior performance to the previous one, it leads to more complicated implementations and also raises concurrency considerations.

8.7.1 INMS Adaptations

The role of the INMS NIM is to make available the WINMAN functionality to the SMS module. Due to the technology-independent nature of the CaSMIM standard, the only adaptation needed for the various NMS systems concerns the changes in interfaces published by the components that comprise the NMS. The only change in the interface exported by northbound is the limitation of the functionality provided by the SMS. This is because the system may not wish to make available to external users some data regarding the topology of the network provider.

8.8 Southbound Interface Manager

The southbound interface manager (SIM) supporting component is introduced in order to provide the needed adaptation of the WINMAN internal proprietary interfaces to the external standardized ones. This adaptation is needed in order to isolate the WINMAN components of the external interface specification. In the event that WINMAN system selects another standard for communication, the only affected component will be the SIM.

8.8.1 INMS Adaptations

The INMS SIM is a WINMAN component responsible for establishing a data connection between the INMS provisioning manager and inventory manager, and the NMS NIM component for each supported technology. This interface is CORBA-like and CaSMIM compliant [6]; it involves commands that the INMS components might want to see executed on the NMS component or notifications that the NMS must propagate to the INMS.

The INMS SIM supports the CaSMIM interfaces listed in Table 8.2, and the notification support is made through the the *Common::Notification::Reciever* interface. Also in this implementation, the CaSMIM standard coarse grain approach was followed for better performance in the communication with the client systems.

Through this interface, the INMS provisioning manager is able to allocate resources for an SNC and to create, activate, modify, deactivate, and release SNCs. On the other hand, the INMS inventory manager is able to get and set object details or status and get, find, and delete objects defined in each NMS. In the opposite direction, the NMSs propagate notifications to the INMS provisioning manager about network state changes or network errors. Notifications are propagated to the INMS inventory manager whenever a new object is created or deleted.

8.8.2 IP and WDM NMS Adaptations

In order not to restrain the WINMAN system to any proprietary interface between the NML and EML, the WINMAN solution follows the TMF, MTNM NML-EML interface specification [6]. Consequently, the SIM of the IP-NMS and the WDM-NMS supports this specification. As this specification is not yet finalized, it is worth mentioning that we used its May 2001 release.

The MTNM interface specification covers many management fields (configuration, connection, fault, performance, protection, equipment, and security) of the NML-EML communication, not all of which are under the scope of the configuration management area treated in this book. Thus, a selection of the MTNM-specified interfaces and operations were made. Specifically, only the MTNM interfaces depicted in Figure 8.8 were used in the current implementation (Figure 8.9).

Moreover, this specification covers only ATM, SDH, and the WDM technology domains and not IP/MPLS. For that purpose, an adaptation of those interfaces/operations was made through the use of the generic information fields provided by the MTNM and the following adoptions:

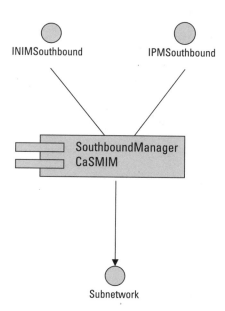

Figure 8.8 The CaSMIM southbound manager context.

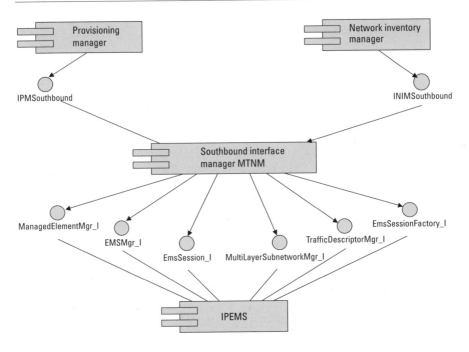

Figure 8.9 The MTNM southbound manager context.

- MPLS-specific information (FEC information that identifies the IP connection end points) is passed through the subnetworkConnection::SNCCreateData_T.additionalCreationInfo field.

- IP QoS information (bandwidth) is communicated through the trafficDescriptor::TDCreateData_T.trafficParameters field.

- MTNM notification format was adapted to cover the IP domain and WINMAN system needs (creation/deletion of IP link).

8.9 The GUI

The primary role of the GUI is to serve as an entry point to the WINMAN system, from which the operator shall have access to the functionality of the NMSs. As already pointed out in Chapter 7, although the WINMAN platform comprises three subsystems (IP-NMS, WDM-NMS, and INMS), the GUI application has a common look and feel for all three and is connected to the desired subsystem. If we want to interact visually with the three NMSs simultaneously, a GUI instance should be launched for each one of them.

The GUI pursues a thin-client architectural approach. The GUI is mainly assigned the task of ensuring the comprehensible visualization of data, and it transfers as much application logic as possible to the lower layers of the system (e.g., the view manager). Nonetheless, it is inescapable that a certain amount of application data and metadata are kept in the GUI. The data corresponds with the entities that are manipulated in the WINMAN system and must be surfaced onto the visual interface, such as network devices, termination points, links, or connections. Procedures such as filtering, or grouping according to certain attributes, are hastened when the data is stored locally. In addition, the metadata is saved in local files, and it contains information that concerns the semantics of the actual data; for instance, the keywords that have been agreed in communication with the view manager.

It was shown in the previous chapter that the WINMAN front end is structured according to the MVC design pattern. Moreover, the MVC paradigm is trailed within the bounds of the GUI component. This is accomplished by separating the classes that contain the data from those intended to visualize it; to make it more tangible, for each of the managed elements, there are two relevant GUI classes—the view class, which has properties and methods that determine the drawing of the element on the canvas, and the model class, which carries all of the significant information and methods to retrieve or alter it.

In detail, the WINMAN GUI supports:

- *Both physical and logical views of the network.* These views are disjoined in the network representation. Physical elements (e.g., OXCs and routers), equipment, cards, and topological links are shown in the physical view, whereas the PTPs and the existing connections are depicted in the logical view. Furthermore, the WINMAN GUI supports further discrimination of the logical views with technological domain criteria; that is, it gives access to five different logical network panels:

  1. An integrated panel showing the IP points that are accessible from the service layer;

  2. A summary IP view, which contains the border PTPs of the IP network;

  3. A detailed IP view, which includes the total IP PTPs;

  4. A summary WDM view, which contains the border PTPs of the WDM network;

  5. A detailed WDM view, which includes all the WDM PTPs.

  The network nodes in each panel are positioned on the plane in an orderly manner forming network clouds with clear outlines. Also, the PTPs belonging to the same physical device are grouped in close proximity to each other in order to exhibit the property of inherence to the corresponding device.

On the left-hand-side of the GUI window of Figure 8.10, a tree gives reference to the various views of the application, in an explorer-like style. As such, whenever a specific branch of the tree is selected, the respective network view is depicted on the right panel.

- *Establishment and maintenance of end-to-end connectivity services.* Using the WINMAN GUI, the user can create, modify, and remove end-to-end connectivity services. These operations are facilitated by user-friendly dialogues, in which certain parameters are specified before sending the request. A sample dialog that requests an ICS is depicted in Figure 8.11. The operator can either define only the end points of a circuit (automatic provisioning) or also define its intermediate points (manual provisioning).

- *Display of attributes of a managed resource.* By selecting a device, a termination point, a link, or a whole connection, the WINMAN operator can ask for the properties that characterize the specific entity, which are then presented in a tabular form.

The GUI communicates with the rest of the system via the view manager component. The view manager is the only point of access towards the

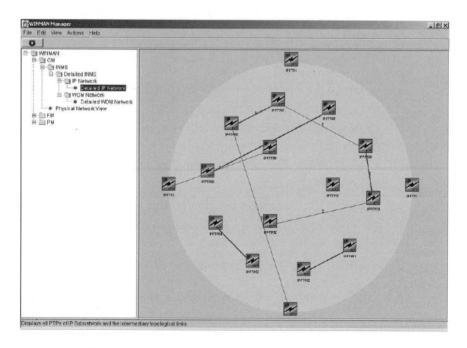

Figure 8.10 A GUI snapshot.

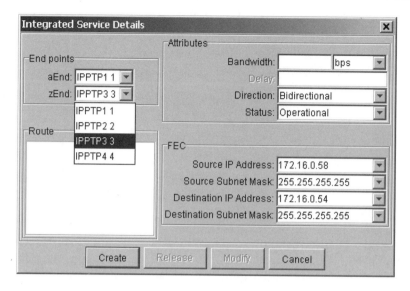

Figure 8.11 Sample of an integrated connection request dialogue.

WINMAN components in order for the GUI client to be as thin, independent, and modular as possible. As a design principle, it has been decided that the data retrieval by the GUI is accomplished in *pull* mode. Therefore, the view manager provides an interface to the GUI, but not vice versa, because in general the GUI can be *mounted* anywhere in a distributed environment. The rest of the components do not need to be aware of its existence and exact location. In order for the GUI client to obtain information, it uses the methods specified by the interface provided by the view manager. To avoid drawing enormous amounts of data in one call (which reduces the performance of the system), iteration methods have been introduced that fragment the large bulk of data into smaller pieces and facilitate unhindered communication.

The object model adopted by the GUI follows a flat structure; that is, for reasons of scalability, a single class: the WinmanObject class, abstracts all of the particular entities that lie in the inventory. The task of translating the nested hierarchical structure of the system to the swallow model of the GUI level is taken on by the view manager server. The WinmanObject class is outlined in Figure 8.12.

Besides its interaction with the view manager, it was intended that the GUI act as a receiver of certain notifications that are related to changes in the status of connections (e.g., it should be notified whenever one connection is created or removed). In such a manner, the GUI depicts in real time an exact picture of the underlying network.

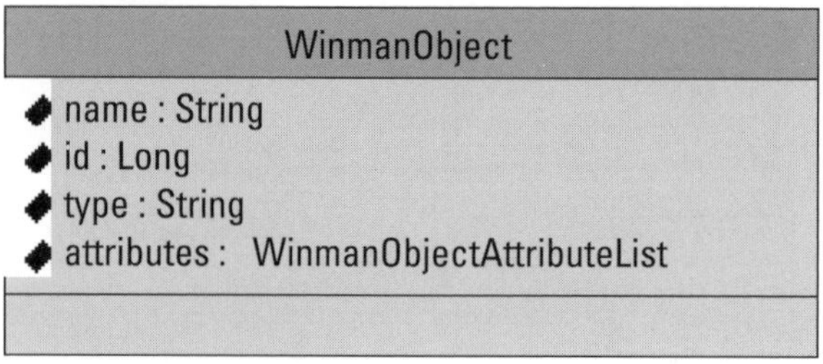

WinmanObject
◆ name : String ◆ id : Long ◆ type : String ◆ attributes : WinmanObjectAttributeList

Figure 8.12 The WinmanObject class.

8.10　Summary

This chapter has described the design of the generic network management system in WINMAN that decomposes into a set of eight components. Additionally, in the description of each component, we have detailed the adaptations that are required to use the component in the specific instances of the WINMAN NMS (i.e., the INMS, IP-NMS, and WDM-NMS).

The main functionality of the provisioning manager is to control the provisioning of connectivity services in the WINMAN-managed IP over WDM network. This component receives the requests issued from the WINMAN operator through the GUI or from the SMS application through the northbound interface and triggers, via the southbound interface, the required configuration actions on lower level applications to activate the services.

The end-to-end routing manager is in charge of calculating routes between two given end points of a requested connection, taking into account QoS and policy constraints.

The network inventory manager is designed to have a full view of the network. Its responsibility is to know the different elements of the network, their roles, and their connections to other elements. The inventory component should have a full view of all of the subnetworks that the configuration module is managing.

The policy manager has two roles. First, it is the WINMAN component responsible for the execution of actions derived from policies to be enforced in the managed network. Specifically, authorization policies and routing policies are considered. On the other hand, it is responsible for the management of the life cycle of these policies.

The view manager is a gateway component that links the GUI to the WINMAN server. Its main role is to enable the GUI to operate independently and to view the entire WINMAN server as only one component. In this way, the GUI has one point of contact to the WINMAN server, regardless of the required functionality or the viewed domain.

The primary role of the NIM is to provide a single entry point to clients that wish to gain access to the services provided by the corresponding NMS in a way that is independent of the underlying technologies. These requirements are in line with the concept of the TMF CaSMIM standard, where it is supported.

The SIM is intended to provide the necessary adaptation of the WINMAN internal proprietary interfaces to the external standardized ones. This adaptation is needed in order to isolate the WINMAN components of the external interface specification. The protocol that is supported for this interface implementation between the IP-NMS and the WDM-NMS with their respective EMSs is TMF MTNM, and the interface between the INMS and the IP-NMS and the WDM-NMS is the TMF CaSMIM.

Finally, the GUI is to serve as an entry point to the WINMAN system, from which the operator shall have access to the functionality of all of the NMSs. Although the WINMAN platform comprises three subsystems (IP-NMS, WDM-NMS, and INMS), the GUI application has a common look and feel for all three and is connected to the desired subsystem.

References

[1]   Raptis, L., et al, "Integrated Management of IP over Optical Transport Networks," *Proc. IEEE International Conference on Telecommunications,* ICT 2001, June 4–7, 2001, pp. 172–177.

[2]   TMF 508,605, and 807, "Connection and Service Management Information Model (CaSMIM)," Public Evaluation, Version 1.5, June 2001.

[3]   Dijkstra, E. W., "A Note on Two Problems in Connection with Graphs," *Numerical Mathematics,* 1959, pp. 1:269–271.

[4]   Vardalachos, Nikolaos, et al., "IP over WDM Networks: A Policy-Based Management," Eighth IFIP/IEEE Network Operations and Management Symposium, Florence, Italy, April 15–19, 2002.

[5]   Damianou, N., et al., "Ponder: A Language for Specifying Security and Management Policies for Distributed Systems, V 2.3," Imperial College Research Report DoC 2000/1, October 2000.

[6]   TMF 513, 608, and 814, "Multi-Technology Network Management (MTNM) NML-EML Interface," Public Evaluation, Version 2.0, August 2001.

9

Technologies and Tools

9.1 Introduction

The performance of a management system depends not only on the architectural aspects that guided its design but also on the technologies that are selected for its implementation. These technologies range from operating systems to high-functionality platforms going through different types of languages, persistency, and representation devices. Therefore a well-designed system requires also an appropriate choice of supporting technology to lead to a full-performing product.

We have said on several occasions throughout the book that among the objectives of the WINMAN solution, the performance aspect was excluded. We are only interested in validating our approach in terms of its functional behavior with respect to specific management functional areas and in verifying some of its nonfunctional behavioral characteristics, although none of them will be an index related to the system performance. The main motivation to exclude this important system attribute was simply that the time and resources needed to reach such an objective were not available in the project. Nevertheless, we wanted to use the design phase of the project to carry out a survey on the available technologies and tools that would adhere to our prototypes—at least the necessary conditions to allow them a further evolution for performing systems. This chapter is thus devoted to summarizing the main findings on this area.

Therefore, we describe hereafter the platforms and tools that were used in the WINMAN development process. For each technology or tool, we have a summary description of its main characteristics and advantages and then we show how they were used in the WINMAN.

The chapter is structured around nine sections plus this introduction and a summary. The specific nine sections are each devoted to one of the technology types that can play a role in the system prototypes. These are development platforms, operating systems, programming languages, middleware, interfaces between components, communication protocols, databases, GUIs, and policy-based management tools.

9.2   Development Platforms

9.2.1   Network Management Platforms

Many commercial tools are built to allow the operators and the vendor to create management functionality for their equipment and networks. These tools are built with network management functionality and allow their users to customize it to their needs.

The main features that are included in those platforms are:

- *Southbound interface.* This usually supports SNMP.
- *Fault.* This allows alarm monitoring.
- *Performance.* This allows collecting performance measurements and displays them on an accumulative graph.
- *Configuration and inventory.* This allows topological views, displays alarms on the views, and creates and sends commands to the network elements.
- *Security.* This allows user authentications.

All of those functions can then be customized to support the specific network and equipment. Nevertheless, those platforms are not development tools and have many limitations in designing an open interface for a multitechnology environment.

In WINMAN, we looked for a platform that would support the component-based architecture and that would allow us to define the components needed and their functionalities. Due to their limitations, we did not want to use a network management platform that already included the network management functionality. Rather, we wanted a development platform that would allow us to create the components that we identified and give them the desired functionality.

Two platforms were selected, both using component-based technology. One is the Java-based distributed software component (DSC) platform, and the other is the C++ based NETRAC platform. They were both used in an integrated and distributed way.

9.2.2 The Distributed Software Component Framework

The DSC framework [1] provides the middleware layer that allows components to interact with each other in a distributed environment. It consists of a CORBA object request broker (ORB), a run-time library, which provides additional services that are not available in the ORB, and tools for testing and validation. A summary architecture of the DSC platform is shown in Figure 9.1. A DSC component may support multiple interfaces. Each interface is specified in object management group (OMG) interface definition language (IDL) and implemented as a CORBA object. Components in the DSC framework may, in principle, be implemented in any language for which an IDL mapping exists. The component logic and run-time libraries are currently implemented in Java, so the components can be deployed on a variety of operating systems and hardware platforms, enabling automatic downloading to many end-user terminals. DSC supports the following concepts:

- *Components.* A component in the DSC framework can be seen as a group of CORBA objects.

- *Nested components.* The DSC framework also supports nested components where one (compound) component contains other

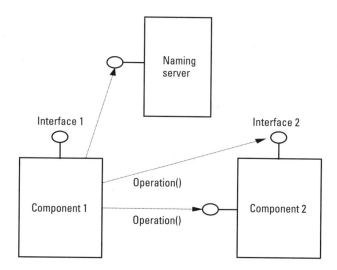

Figure 9.1 Architecture of the DSC platform.

(sub)components. It allows component specialization through aggregation rather than inheritance.

- *Component containers.* Each component in the DSC framework belongs to a component container. The container is a specialized compound component that provides the run-time context in which components operate.

The DSC framework provides an environment, allowing the testing of individual operations on component interfaces and the generation of a complete sequence diagram as an invocation propagating through the system. The test system tracks invocations between component interfaces and generates all of the information needed to identify the source and destination interface, the operation, and the parameter values. This information can be collected to generate a complete message-sequence diagram for an invocation, which can then be displayed graphically. During system verification, the message-sequence diagrams that are generated may be compared to the diagrams that were created during the system specification phase.

9.2.3   The Netrac Platform

The Netrac platform[1] is designed for developing full applications that have a multitier architecture. Therefore the platform is combined from three tiers:

- The server tier, consisting of a component-based architecture. It is written in C++.
- The middle tier, which handles the presentation logic and the object model. It is based on the ILOG TGF product [2].
- The client tier, consisting of a plug and play GUI. It is written in JAVA 1.3.

Netrac platform is designed to support easy development and easy on-site customization. Figure 9.2 shows its high-level architecture.

The main characteristics of the platform are listed as follows:

- The application behavior is defined by connecting different closed components with a scripting language.

---

1. The Netrac platform is embedded in the commercial solutions of TTI Telecom and is not sold as a standalone system.

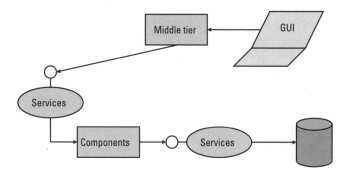

Figure 9.2 Architecture of the Netrac platform.

- The application behavior can be customized on site using scripting language and without recompiling the code.
- The client is Web enabled and can be run using any browser.
- The presentation is metadata driven and therefore can be updated to meet the needs of the customer.
- The system supports multilanguages display by updating the presentation metadata.
- The system has a plug and play mechanism, which enables plugging of new components for server behavior and plugging of GUI widgets and control for changing the client view.
- The system can integrate third-party products in its components, such as ILOG libraries.
- The system is platform independent. It can run on NT 4.x and Windows 2000, SUN SOLARIS 2.6, HP 11, IBM AIX 4.3, and others.
- The system is database independent. It can work with Sybase, Oracle, and others.

9.2.4 The WINMAN Implementation

The WINMAN developers used the platform as the basic structure of their component. In both platforms, the WINMAN components were built as an implementation and expansion of the platform components. In this way, all of the basic functionalities that the platform components support were used. The component interfaces and their data structures were platform based, and the functionality and exposed operations were developed on WINMAN.

As mentioned earlier, the WINMAN solution consists of two platforms, a Netrac C++ platform and a Java DSC framework. In order to connect these two platforms, a gateway was developed that connects both platforms. XML was chosen as the intermediate language between them. The gateway consists of two parts, a Netrac part adapting the Netrac internal format to XML and a DSC part, which converts CORBA IDL messages to XML messages. The gateway supports all CORBA concepts like return values, in/out parameters, exceptions, and notifications. In the way it is implemented, the Netrac part plus the gateway can be seen as a normal DSC component. Seen from the Netrac point of view, there is only one platform with a connection to the outside world.

9.3   Operating Systems

Operating systems provide a software platform on top of which other programs, called application programs, can run. Operating systems perform basic tasks. For large systems, the operating system makes sure that different programs and users running at the same time do not interfere with each other. It is also responsible for security, ensuring that unauthorized users do not access the system. Therefore, the choice of the operating system determines to a great extent the applications that can be run.

9.3.1   UNIX

UNIX is an operating system that was created in the late 1960s in an effort to provide a multiuser, multitasking system. The philosophy behind the design of UNIX was to provide simple, yet powerful, utilities that could be assembled together in a flexible manner to perform a wide variety of tasks.

The features that have contributed to UNIX success are:

- *Multitasking capability.* Several applications or processes can be run at the same time.
- *Multiuser capability.* UNIX allows several users to share the use of the computer at the same time and with different profiles.
- *Portability.* It is programmed in C, the most portable operating system in history.
- *Initially distributed as source code.* This allowed many programmers to contribute to its development.
- *Integrated.* The UNIX programming environment interconnects thousands of tools in an easy way.

9.3.2   Windows

Windows 2000 Professional is the windows operating system for business desktop and laptop systems. It is used to run software applications, connect to the

Internet and intranet sites, and access files, printers, and network resources. Windows 2000 is designed to appeal to small business and professional users as well as to the more technical and larger business market for which the NT was designed.

9.3.3 The WINMAN Implementation

In the WINMAN implementation, each partner used the operating system that he or she preferred. The integration of the software was done on a Windows 2000 operating system due to the simplicity of the operating system and to the fact that it is portable and could be installed and operated using a laptop.

The WINMAN consortium decided to follow the trend in commercial network management applications and made a UNIX implementation of the system. This decision will significantly ease the exploitation plans of commercial partners, as the solution adopted resembles very closely what a network operator would actually deploy as a management system.

Therefore, the WINMAN solution is also ported completely to UNIX and can be used on both operating systems.

9.4 Programming Languages

Object-oriented programming (OOP) is organized around objects rather than actions and data rather than logic. OOP allows the description of the problem in terms of the problem, instead of in terms of the computer where the solution will run.

9.4.1 Java Programming Language

Java is twofold: a programming language and a platform. From the point of view of a high-level programming language, it exhibits the following characteristics:

- Simple;
- Object oriented;
- Distributed;
- Interpreted;
- Robust;
- Secure;
- Architecture neutral;
- Portable;
- High performance;

- Multithreaded;
- Dynamic.

Each Java program is both compiled and interpreted. With a compiler, the Java program is translated into an intermediate language called *Java byte codes*—the platform-independent codes interpreted by the Java interpreter. With an interpreter, each Java byte-code instruction is parsed and run on the computer. Figure 9.3 represents the portability concept exhibited by Java.

The Java platform has two components: the *Java Virtual Machine* and the *Java Application Programming Interface* (Java API).

9.4.2   C++ Programming Language

The C++ programming language provides a model of memory and computation that closely matches that of most computers. In addition, it provides powerful and flexible mechanisms for abstraction; that is, language constructs allow the programmer to introduce and use new types of objects that match the concepts of an application. Thus, C++ supports styles of programming that rely on fairly direct manipulation of hardware resources. In this way, it is possible to deliver a high degree of efficiency.

9.4.3   The WINMAN Implementation

WINMAN was developed using both programming languages. The DSC platform is a JAVA-based platform and the Netrac platform is a C++-based

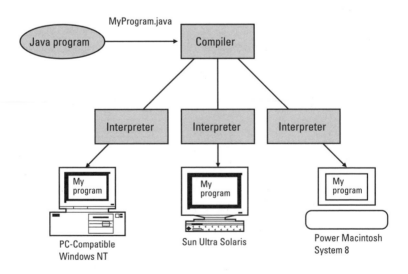

Figure 9.3 The Java code portability concept used in WINMAN.

platform. The DSC gateway was developed in Java and the Netrac getaway was developed in C++. Java, being a platform independent language, can run on a Windows 2000 and a UNIX operating system. The C++ code developed on the Netrac platform was ported from the Windows 2000 environment to the UNIX operating system, thus allowing the entire WINMAN project to run on both environments.

9.5 The Middleware

The middleware in the context of WINMAN is connectivity software consisting of a set of enabling services that allow multiple processes running on one or more machines to interact across a network. It is a general term for any programming that serves to "glue together" or mediate between two separate and usually already existing programs.

By introducing middleware in WINMAN, it is possible to separate responsibilities between parties and to shield machine and operating system dependencies. The middleware used in WINMAN glues together the management applications with a standard Java or C++ platform by providing:

- Global communication services;
- Naming and location services;
- Notification service;
- Component support.

9.5.1 CORBA

CORBA is a technology for distributed object computing defined by the OMG, a cross-industry consortium. It defines common semantics for the specification of objects and the interfaces they offer, allowing standard and programming language–independent communications between applications.

The main components of the CORBA standard are:

- The ORB that provides a platform allowing interobject communication, without dependency on the specific platforms and techniques used to implement the objects. The ORB lets the objects send and receive requests and responses in a transparent way within a distributed environment.
- The CORBA services, which implement generic and commonly used functions through standard interfaces. The name service, the event service, and the lifecycle service are examples of already defined services.

- The common facilities that provide a set of application functions that can be configured to satisfy the requirements of a specific configuration, providing vertical solution in certain business areas.

Table 9.1 shows the most important additions included in the successive versions of CORBA.

9.5.1.1   Advantages of CORBA

The following benefits of the use of CORBA for the development of management systems have been identified:

- It offers a standard and well-known mechanism for communication between any kind of application. In addition, CORBA-based platforms are becoming more interoperable. This, together with the implementation-technology independence, helps applications become part of the distributed computing environment.
- It allows for component reuse. Small components, with a clearly defined behavior and compliant to a specified interface, can be built and easily introduced in systems that use CORBA as the distributed computing technology, no matter what operating system or programming language has been used.

Table 9.1
The Versions of CORBA

CORBA Version	Added Features
Version 2.0	The Internet inter-ORB protocol (IIOP) specification, allowing a CORBA ORB from one vendor to communicate with an ORB from another vendor
	IDL language mappings for C++
Version 2.2	The portable object adapter (POA), allowing object implementations portability between different ORB products
	IDL language mappings for Java
Version 2.3	The objects-by-value feature—WINMAN choice
Version 2.4	Real-time CORBA
	Asynchronous messaging
Version 3.0	Java-to-IDL mapping
	Firewall support
	CORBA component model

Joint interdomain management (JIDM) defines how network management components based on OSI and SNMP can interoperate with CORBA-based components. Hence, the interfaces and information models already developed following CMIP and SNMP standards can be used to build a CORBA-enabled interface wrapping the existing application, taking advantage of CORBA characteristics.

The DSC platform used in WINMAN is a lightweight implementation of the CORBA component model, providing the elementary services for building component-based applications.

9.5.2 XML

XML is a metamarkup language that provides a format for describing structured data. Table 9.2 presents the main characteristics of XML and the areas where it can play its role.

Some of the main concepts related to XML are:

- *XML documents.* These contain one or more *elements* delimited by start tags and end tags. These elements maintain a hierarchical relationship. An XML document is considered well formed if it contains exactly one root element (the document), and all the children are nested properly within each other. That is, both begin and end tags of a given element should exist within the body of the same parent element.

- *Document type definition (DTD).* This is a set of rules that a document follows that software may need to read before processing and displaying a document.

- *Schema.* This is an XML-based syntax for defining how an XML document is marked up. DTDs have many drawbacks, including the use of non-XML syntax, no support for data typing, and nonextensibility. XML schema addresses all of these problems.

Table 9.2
Characteristics of XML and Its Areas of Application

Key Attributes of XML	Key Areas of XML
Simplicity	Data interchange
Extensibility	Data integration
Supports hierarchical information description	Document management/publishing
Supports linking	Formatting/presentation

- *Style sheets.* These describe how documents are presented on screens, in print, or perhaps how they are pronounced.

- *XML parsers.* An XML parser (also known as XML processor or XML API) is a software module capable of reading XML documents and providing access to their content and structure. Information in XML documents, which are stored in some kind of persistence engine, is accessed through the services of a Java/C/C++ XML parser. There are two main API specifications that have gained popularity among developers today and are striving to become industry standards: the document object model (DOM) and the simple API for XML (SAX).

- *Simple object access protocol (SOAP).* This is a lightweight protocol for the exchange of information in a distributed environment. It consists of three parts: an envelope that defines a framework for describing what is in a message and how to process it, a set of encoding rules for expressing instances of application-defined data types, and a convention for representing remote procedure calls and responses. SOAP can potentially be used in combination with a variety of other protocols, like HTTP and SMTP.

9.5.2.1 Advantages of XML

The primary advantages of going in for an XML solution are the following:

- *Development of flexible applications.* XML, together with HTML for display, scripting for logic, and a common object model for interacting with the data and display, provides the technologies needed for flexible three-tier application development.

- *Data integration from disparate sources.*

- *Local computation and manipulation of data.*

- *Multiple views of the data.* XML complements HTML. While HTML describes the appearance of data, XML describes data itself. As display is now separate from data, having this data defined in XML allows different views to be specified, resulting in appropriate data representations.

- *Granular updates.* Only the changed element must be sent from the server to the client, and the changed data can be displayed without refreshing the entire user interface. Presently, an entire page must be rebuilt if even one item of data changes, even when the view remains constant. This severely limits server scalability.

9.5.3   The WINMAN Implementation

The WINMAN used two types of middleware—one between its components and the other one between the platforms and between the GUI and the server.

- *Components middleware.* The components used the CORBA IDLs to communicate and transform data objects.

- *Platform middleware.* The platforms used their native middleware. The DSC platform used the CORBA IDLs and the Netrac used the XML. The DSC gateway translated the IDLs into a descriptive XML format that was read in the Netrac gateway and transformed to its native data structure.

- *Client-server middleware.* The client-server communication used XML to describe the requests and the data objects.

9.6   Interfaces Specification

This section describes the tools supporting the interfaces between the WINMAN NMSs and the EMSs.

9.6.1   CORBA

Besides what has been already exposed, CORBA specifies a standard for implementing and accessing ORBs via common interfaces, specified in OMG's IDL. The following requirements are taken into account:

- Interface definition is to be separated from its implementation.

- Interface typing (object references are typed by interfaces) and inheritance are to be used for the extension, evolution, and specialization of the objects behavior.

- An ORB manages the requests of the client objects against the services provided by server objects. The ORB is mainly intended to provide for interoperability.

The usage of CORBA is not limited to the NMS/EMS interface, which is addressed by WINMAN, but is also extended into the EMS and into the network elements of the WDM network. In this way, experience could be gained about the usage of CORBA on all management layers.

9.6.2 SNMP

The simplicity of SNMP has been a major factor in its success leading to a rapid adoption by manufacturers and a huge base of agreed MIB definitions. There are a number of sophisticated products such as HP OpenView (e.g., network node manager), which offer powerful facilities for the management of SNMP-based networks. From a user's point of view, SNMP offers the benefit of wide industry support and comparative cheapness. However, SNMP suffers from a number of deficiencies:

- The protocol is very inefficient when large quantities of information have to be retrieved from a managed system.
- There is poor support for event-driven management.
- There is, in effect, no security.

The last point is the most serious one, and it reduces SNMP to a monitoring rather than a management protocol. SNMPv2 was intended to remedy the principal deficiencies of SNMP and be largely backwards compatible with it. Unfortunately the adoption of SNMPv2 and later versions has not been widespread.

9.6.3 The WINMAN Implementation

The WINMAN interface specification is written entirely in CORBA IDL and the component-specification language (CSL) used by the DSC framework. The CSL is based on the CORBA component model specification.

The DSC CSL is the language used to describe the components that client objects call and object implementations provide. A CSL component provides the information needed to develop clients that use the component's operations. Clients are not written in CSL, which is purely a descriptive language, but in languages for which mappings from CSL concepts have been defined—for the WINMAN implementation, this is Java.

The CSL can specify static compound components and list per subcomponent which of its interfaces are to be exported. Per component, it lists generated and accepted notifications, exported and required interfaces, and component properties. Also, the CSL lists per interface generated and accepted notifications and interface-specific properties. Supported interfaces can be specified to be dynamic or static.

9.7 Connectivity Interface Protocols

There is a broad spectrum of communication protocols available for conveying management information. One of the oldest is SNMP. As already pointed out, the simplicity of SNMP has been a major factor in its success but this same simplicity was also one of the reasons for needing a more robust management protocol. Therefore, a protocol stack based on CMIP and common management information service element (CMISE) was specified. Although these protocols provide a good method for describing the management information, there was no large penetration of the market. A good alternative between these extremes is CORBA. It provides both a safe environment and is far less complex then CMIP/CMISE.

Alternatives like XML (which is seen by some as the successor for CORBA IDL) were not taken into account, as these alternatives are not yet mature and still have to prove their value as management protocols.

9.7.1 Connection and Service Management Information Model

Today's telecommunications world is dealing with all types of services. This situation has led to a business approach of a quick and easy service creation, taking into account the impact of network faults on services and having visibility of the problems on several layers. These goals led to the creation of the CaSMIM approach. The CaSMIM goal is to create a standard model to automate service provisioning and end-to-end monitoring across multitechnology, multivendor, and multilayered networks. This involves translation of service requests *issued by the* SMS into network connection provisioning actions. Another goal is to be *service oriented* but *technology independent,* and so be flexible to cater for connectivity setup across networks using current and future technologies. Figure 9.4 shows the CaSMIM protocol between network layers.

CaSMIM sees its usage in the following scenarios:

- *Service provisioning.* This is for automatic provisioning of service whatever the underlying technology implied.
- *Service assurance.* This is to be used to export the visibility of a given network layer and to show the impact of a given network failure on the service.

9.7.2 Multitechnology Network Management

The objective of the MTNM specification is to detail, with the usage of UML, the interface between entities that reside in the NML with entities of the EML. The successful fulfillment of the above goal provides for a scalable and nonproprietary network management solution, where multivendor, multitechnology (ATM, SONET/SDH, WDM) management systems interoperate in an open architecture. Examples of the management business scenarios that the MTNM

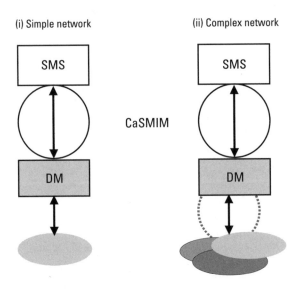

Figure 9.4 The CaSMIM protocol between network layers.

specification caters for are: inventory discovery, connection provisioning, equipment provisioning, and performance management.

The MTNM specification consists of the documents referenced as [3–5]. The [5] contains an implementation of the specified interface using the IDL notation of CORBA, while [3] defines the problem, requirements and use cases that the MTNM specification solves and [4] contains the information model that is used in order to represent the managing network.

9.7.3   The WINMAN Implementation

To achieve vendor- and technology-independent NML-EML communication, the WINMAN system followed the MTNM specification. Although the adoption of this specification means that the WDM technology domain is covered, as it is inherently supported by the MTNM, the WINMAN system also requires IP domain communication functionality. This problem was solved, in a consistent manner to the specification, utilizing the existing generic fields and operations in order to pass IP-specific information, without requiring any major modifications. This successful application of the MTNM specification to the IP technology domain, which was not considered by the TMF, qualifies the correctness of our choice.

The CaSMIM interface was used for implementing the interfaces between the IP NMS and the WDM NMS to the INMS and further between the INMS and the SMS. The WINMAN NMSs northbound component implemented the

CaSMIM protocol and thus created a public interface between the WINMAN system and the SMS.

9.8 Databases

When associations among files or records cannot be expressed by links, relational databases are used; a simple flat list becomes one row of a table, or *relation*, and multiple relations can be mathematically associated to yield desired information. The user must have an understanding of how data is structured within the database in order to retrieve, insert, update, or delete records. However, it is much simpler to understand than hierarchical models, and it provides extremely useful tools for database administration.

On the other hand, object-oriented databases store and manipulate more complex data structures, called *objects*, which are organized into hierarchical classes that may inherit properties from higher classes in the chain; this database structure is the most flexible and adaptable.

9.8.1 The WINMAN Implementation

WINMAN used a relational database because this is the state of the art as used in the network management industry. Because our software used an object-oriented data model, we had to do an object-to-relational mapping. The methodology we used is the simplest that maps each object to a table, each structure or list to a table, and each parent/abstract object to a table. This was done to enable the retrieval of the basic (and most often queried) information stored in the abstract objects in one database query.

9.9 User Interface Tools

The term *user interface* refers to the methods and devices that are used to accommodate interaction between machines and the human beings who use them (users). Among the properties of the GUIs are intuitiveness, consistency, simplicity, prevention, forgiveness, and aesthetics.

9.9.1 Hypertext Markup Language

HTML was introduced in the mid 1980s as a subset of standard generalized markup language (SGML). SGML was a metalanguage defined by IBM in the early 1970s as a method for describing text and documents that were not specific to any device or application. Based on this concept, W3C originally introduced HTML as a presentation mechanism for hypertext documents exchanged between members of the scientific community of the Internet. Combined with the HTTP protocol that had the role of the transport mechanism, HTML was widely accepted as an emerging standard for data presentation in the new World Wide Web (WWW) network. Very soon the popularity, ease of usage, and the abundance of HTML authoring tools convinced the users to adopt HTML for performing a wide new range of roles. New roles meant new requirements and most notably the demand for delivering dynamic content to the user. It was clear that for this task, the classic implementation of an HTTP server sending static HTML pages to client devices was no longer adequate. The new applications could no longer be based on the traditional client-server paradigm. This led to the emergence of new architectures that aimed at covering the new needs. One such methodology was the three-tier architecture that is followed for the WINMAN implementation. According to this, the system consists of three layers: the back end, the business layer, and the presentation layer.

9.9.2   Java

From the early years, Java provided a rich set of libraries and tools aimed to help programmers in building complex user interfaces. An early implementation of such a library was AWT [6], which was provided with JDK 1.0 and 1.1 platforms. While AWT provided to the user an easy way to utilize a rich set of graphic interface elements, it also suffered from several major disadvantages. Most notably, AWT components were implemented using libraries that were system dependent. This meant that the AWT-based applications were restricted to lowest common denominator of the features provided by the system-dependant libraries.

The Java foundation classes (JFC) were introduced in 1997 to overcome this problem. The JFC contained the following:

- *The swing components.* These include everything from buttons to split panes to tables.

- *Pluggable look and feel support.* This gives any program that uses swing components a choice of looks and feels (e.g., the same program can use either the Java look and feel or the Windows look and feel). We expect many more look-and-feel packages—including some that use sound instead of a visual look—to become available from various sources.

- *Accessibility API.* This enables assistive technologies such as screen readers and Braille displays to get information from the user interface.

- *Java 2D^{TM} API (Java 2 platform only).* This enables developers to easily incorporate high-quality two-dimensional graphics, text, and images in applications and in applets.

- *Drag and drop support (Java 2 platform only)*. This provides the ability to drag and drop between a Java application and a native application.

The first three JFC features were implemented without any native code, relying only on the API defined in JDK 1.1. As a result, they could and did become available as an extension to JDK 1.1. This extension was released as JFC 1.1, which is sometimes called *the swing release*. The API in JFC 1.1 is often called *the swing API*. Although the Java 2 platform still supports the AWT components, programmers are strongly encouraged to use swing components instead.

9.9.3 · The WINMAN Implementation

For the implementation of the GUI, Java is utilized. The undisputable characteristics that render Java as the programming language of choice are well known: platform portability, underlying technology independence, security, and competence in GUI development. Only pure, open-source Java has been adopted for the graphical interface in order to be compliant with the established Java standards and ensure the portability and interoperability of the GUI application. The version of Java used is the standard edition SDK 1.3.

All of the visual objects that comprise the WINMAN GUI are taken from the swing package, which provides lightweight components; that is, they separate the model from the view of a graphic class. In such a manner, the graphical representation of a given class can be altered without affecting the data and the logic of the associated model and there can be multiple views of the same object. For instance, a network element can be depicted either as a node in the network tree or an image on the network topology canvas.

Apart from the standard visualization classes of Java SDK, helper classes are introduced that facilitate the integration with the DSC framework described in Section 9.2.2. As such, the GUI is handled as a common DSC component. Its mere difference from the rest of the components is that it does not publish an IDL interface.

9.10 Policy-Based Management Tools

Policy development tools are intended to create, verify, and implement policies to be used in policy-based management systems. In some respects, these tools are similar to correlation tools, but in this case they are oriented to the most specific aspects of network management. Thus, ILOG Rules, which is a rule-based system, could be used.

Regarding free/open source, Table 9.3 is a representative sample to be considered.

9.10.1 Ponder

Ponder [7] is a declarative and object-oriented language for specifying policies, providing a common means of specifying policies that map onto various access-control implementation mechanisms for firewalls, operating systems, databases, and Java. It supports policies that are event-triggered condition-action rules for policy-based management of networks and distributed systems. As Ponder is declarative, strongly typed, and object oriented, the language flexible, extensible, and adaptable to a wide range of management requirements.

Ponder has four basic policy types: authorizations, obligations, refrains, and delegations and also has three composite policy types: roles, relationships, and management structures that are used to compose policies. Ponder also has a number of supporting abstractions that are used to define policies: domains for hierarchically grouping managed objects, events for triggering obligation policies, and constraints for controlling the enforcement of policies at runtime.

9.10.2   The WINMAN Implementation

The WINMAN used both Ponder and ILOG Rules for implementing its policy-based system. The Ponder tool was used for the implementation of authorization and obligation policies. Authorization policies support the provisioning at INMS, IP-NMS, and WDM-NMS levels, and the obligation policies help the WINMAN operator to trigger routing actions in the INMS, IP-NMS, and WDM-NMS routing components. These actions can be executed automatically or programmed to be executed at certain times introduced by the WINMAN operator. In the performance management area, policies were used

Table 9.3
Sample of Policy Management Tools and Protocols

Policy Management Protocols	Policy Management Tools
LDAP	Intel COPS-SDK
SNMP	Vovida COPS implementation
Remote authentication dial-in user service (RADIUS)	Lulea University COPS implementation
COPS in the UCLA's implementation	Iphighway COPS-SDK
DIAMETER protocol	

for thresholds configuration, performance data collection configuration, threshold crossing events, and QoS parameters configuration.

The ILOG Rule tool was used for collecting the alarms and finding their root causes. In the same process, it was used for finding the service impacts of the alarms and creating new alarms for the affected subnetwork connections.

9.11 Summary

The WINMAN project aimed to develop an integrated solution for interdomain network management. In order to create an open and standardized solution, a component-based system architecture was selected.

A lot of effort was spent selecting the most appropriate tools for such system development. Moreover, in many cases, more than one tool was selected to demonstrate and suggest more than one development option.

In the operating systems area, the system was developed on a Windows 2000 operating system, which proved to be more simple and *portable* than other options. It was ported also to a UNIX environment.

For the development language, the system development was done on two object-oriented languages: Java and C++. Both languages proved to be valid solutions, and our suggestion is to use the one that is more comfortable for the development team.

As far as the middleware, two middleware tools were used for communicating between the system components, CORBA and XML. XML proved to be more helpful in describing the object, while CORBA was more helpful in mapping the operations. Therefore our recommendation is to use XML only as a descriptive language and to use the CORBA interfaces for negotiating between the software components.

With respect to databases, the decision was to use a relational database, as it is the standard in the industry to choose a simple freeware database and to support an open interface. In this way, the WINMAN solution is database independent and can be run with any relational database.

For the external interfaces, two protocols were used for communicating between the systems. CaSMIM was used for communicating between the NMSs, and MTNM was used for communicating between the NMS and the EMS. This proved to be the right approach because the CaSMIM is more connection and service oriented and MTNM is more equipment and configuration oriented.

The GUI used Java as its display tool because of constrains that HTML has in creating dynamic objects. The system did not support a Web client; therefore, the right conclusion was to use the Java capabilities in drawing dynamic views and receiving dynamic commands from the operator. All of the visual objects that comprise the WINMAN GUI are taken from the swing package, which provides lightweight components (i.e., they separate the model from the view of a graphic class). In such a manner, the graphical representation of a given class can be altered without affecting the data and the logic of the associated model and there can be multiple views of the same object.

Policy and correlation tools were selected, allowing WINMAN to demonstrate the usage of policy-based decision making in the solution. The Ponder tool was used as an integrated Java component in the routing mechanism, QoS thresholds mechanism, and alarm correlations.

To conclude, the WINMAN solution recommendation is to use one of the suggested tools to evaluate the needs of the system and the development team in order to select the most appropriate tool for the task.

References

[1]   Meeuwissen, H., H. Batteram, and J. Bakker, "The FRIENDS Platform—A Software Platform for Advanced Services and Applications," *Bell Labs Technical Journal,* July–September 2000, pp. 59–75.

[2]   *ILOG User's Guide.*

[3]   TMF 513, "Multi-Technology Network Management Business Agreement," September 2002.

[4]   TMF 608, "Multi-Technology Network Management Information Agreement," August 2002.

[5]   TMF 814, "Multi-Technology Network Management Solution Set," August 2002.

[6]   The AWT Java Foundation Class description, java.sun.com/products/jdk/awt.

[7]   PONDER Policy Specification Language, www.dse.doc.ic.ac.uk/Research/policies.

10

Testbed Setup for the Evaluation of IP over WDM Management Systems

10.1 Introduction

One of the most important issues while building an NMS is the validation of the solution on a testbed as close as possible to real network environments. Compared to the validation-employing simulation methods, the use of real network testbeds has several advantages. Among them, we can mention the higher reliability of the validation results, the need to face compatibility issues among different vendor products, and the development of the necessary middleware for adaptation purposes that will improve the know-how and pave the way for future deployment of the NMS. On the contrary, one of the drawbacks of this approach is the cost of the infrastructure. As the consortium that has been developing the WINMAN solution already had the necessary building blocks of a real testbed, this alternative was adopted for the validation process. Nevertheless, once the basic building blocks were available, the task was not just to plug and play. In fact, it was necessary to set up the environment and to tune the hardware and software infrastructure to the specific conditions, so that it was able to fulfill its main goal satisfactorily—to determine the compliance of the system under test with the requirements that were established at the design phase. We devote this chapter to describing this process, which led finally to the WINMAN testbed.

As we will see, the WINMAN testbed is made up of the communications infrastructure—the IP, the WDM, and any other technology subnetworks, their corresponding EMSs, and also a set of monitoring and auxiliary tools needed for the operation of all these components. Henceforth, the NMSs are not considered part of the testbed because they constitute the WINMAN management system that is the subject of the evaluation.

This chapter is structured in five sections. In Section 10.2, we present the functional and high-level requirements to be fulfilled. These requirements apply to four different areas—requirements on testbed management systems, requirements on the validations tools that will be required to deal with both IP and WDM, and finally the requirements of both types of transport network infrastructure. Section 10.3 describes the testbed itself, based on the final configuration adopted by the IP and WDM parts, as well as the integration mechanism used to bring the IP over the WDM. Finally, Section 10.4 outlines the implementation of the adaptation required to interconnect the available EMSs to the WINMAN NMSs. The two remaining sections are this introduction and the summary ending the chapter.

10.2 Testbed Requirements

The testbed requirements are classified according to the different components that will play a role in the validation process. These are the management equipment, the testing and verification tools, and the involved transport network infrastructures. Also, we remind the reader that the term NMS is used in the context of this chapter to refer to the NMS to be tested.

10.2.1 Requirements on the Management Platforms

The general characteristics of the management platforms that the testbed must supply are divided into high-level requirements and functional requirements as follows.

10.2.1.1 High-Level Requirements

High-level requirements are intended to make possible the basic interaction between the NMSs and the network they manage. One fundamental characteristic of the testbed is that it must be connectable to the NMSs. This connection will be performed through the southbound interface of the NMSs under test. For instance, WINMAN defines the interfaces with the IP and WDM EMSs, and the testbed shall be adapted to the WINMAN IP-NMS and WDM-NMS public interfaces. The same principles would apply for other solutions that were different from WINMAN.

The EMS testbed platforms shall apply to the network elements the operations activated by the NMSs. They shall also forward to the NMSs information collected from the network elements resulting from previous operations. Eventually, the agent of a network element may act as a proxy agent of adjacent network elements. The testbed should provide the capability to export data to the NMSs in postprocessing mode (i.e., by using standard SQL commands).

The testbed shall collect management information from the supervised resources and transfer this information in the appropriate protocol format through the EMSs to the NMS applications. The structure of the EMS MIBs shall support the NMS functionality. The MIBs shall be fully open and available for general purpose use (i.e., for implementation through network management development tools). On the other hand, the management information between network elements and their EMSs could be transferred either inband or outband. The transfer mode of the management information shall be transparent to the NMSs.

The testbed network shall operate smoothly, in case its EMSs become disconnected. Also, the testbed shall include equipment from different providers to test multivendor characteristic of the WINMAN solution.

10.2.1.2 Functional Requirements

Functional requirements refer to the functions that the testbed, and particularly the EMSs, have to perform in order to check the management functionality of the NMS prototypes in the three management functional areas for which this has been conceived—to allow configuration, fault, and performance management.

The testbed will allow the NMS to request for information about network resources and also to allow requests on resources status. This will be useful for network discovery. Moreover, as the NMSs must maintain an updated view of the managed network, they should receive notifications of the network topology changes. Therefore, the EMS testbed platforms should send notifications to the NMS with information on the configuration or state changes produced on the network resources.

The information that the NMSs need for alarms processing will be provided by the alarms generated in the network elements. Therefore, the testbed will send alarm notifications to the NMS. Moreover, the testbed should perform correlation and filtering of unsolicited messages coming from the network elements and provide the output of such correlation to the NMSs for further processing. The testbed will allow measuring performance parameters, having the capacity to fix threshold for performance parameters. Whenever these thresholds are reached, the testbed shall send threshold-crossing notifications to the corresponding NMSs. Also, the testbed should be able to keep historical data (log records) and provide them to the NMS upon request.

10.2.2 Requirements on the Testing and Verification Tools

There are many system aspects that can be objectives for testing. Just consider performance, regression, conformance, usability, acceptance, and many more. In some cases manual testing is the most suitable method. There is not one single tool that handles all aspects, but combinations of tools will give a robust testing platform.

General-purpose management platforms are widely used in network management. They have many useful features that could help to troubleshoot or even evaluate the use of WINMAN NMS, such as autodiscovery and graphical display of the environment for easy monitoring and administration, remote administration, and MIB browsing. A general-purpose management platform is desirable to be available in the testbed environment for evaluation and troubleshooting purposes.

10.2.2.1 IP Testing Tools

A traffic analyzer is a mandatory tool for the testbed. It will be connected to the interface of the IP routers to generate both useful data traffic and interfering traffic. It is also required that this tool can monitor specific traffic parameters like at least the flow rate.

The availability of a protocol analyzer would also be desirable, at least at the debugging phase of the interfaces.

10.2.2.2 WDM Testing Tools

Multiwavelength systems introduce an additional complexity in testing and troubleshooting compared to single-wavelength systems. Testing tools are required with characteristics adapted to the much more stringent needs of wavelength division systems. The major new requirement is the need to characterize components and links accurately as a function of the wavelength.

The parameters to be monitored in the WDM layer were already presented in Chapter 4. As we said, most of these parameters can be measured by means of an optical spectrum analyzer. Therefore we consider mandatory this measurement device in the WDM testbed. Besides a spectrum analyzer, other tools can be desirable. Among them, we mention the following:

- *Wavelength meter.* This is required for more accurate characterization of laser sources.

- *Optical loss test sets.* These are used to measure the power in individual channels at the output of demultiplexers, both in the 1,530–1,565 nm and optical supervisory channel band.

- *Optical time domain reflectometer (OTDR).* This is used to determine optical lengths, component and link losses, as well as defect detection.

- *Back reflection meters.* These meters measure the back reflection level on each channel. Important measurement related to back reflection induced by instabilities due to DFB lasers.

10.2.2.3   CORBA Testing Techniques and Tools

CORBA is the technology used for the implementation of the communication between distributed software modules. Figure 10.1 depicts a generic architecture for a CORBA-based NMS managing a domain with three types of underlying EMSs. The communication between the NMS and each EMS is based on CORBA IIOP.

Testing of CORBA implementations can be composed of some of the techniques/tools that are described next:

- *Object-level tester of the behavior of CORBA-based servers.* Usually CORBA object interfaces can only be accessed programmatically. Hence, the availability of a tool that allows the user to connect to multiple servers, view information about live objects within the servers, invoke methods with parameters, and view or modify attributes is desirable.

- *Monitors of distributed communications.* This is an object-tracing and monitoring tool that traces CORBA method calls, monitors outgoing and incoming method calls, and displays requests, replies, and exceptions with their execution times in a GUI.

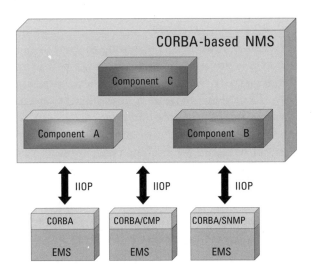

Figure 10.1   CORBA-based NMS/EMS environment.

- *Simulator of real-world CORBA traffic.* This is a tool that simulates thousands of users interacting with servers, which is used to measure capacity and scalability.

10.2.3   Requirements on the IP Infrastructure

- *Network layer requirements.* The basic requirement for the IP routers is that they support IP-v4 with at least a fast Ethernet physical interface. Other types of interfaces like GbE might also be desirable. Among layer-2 protocol availability, the requirement is to have the point-to-point protocol. Other protocols like HDLC, ATM, or SDH—although quite common in commercial routers—are not considered necessary for the testbed.

- *IP layer requirements.* With regard to IP routing protocols and QoS and security support, the network elements of the testbed must support OSPF due to its QoS-based routing capabilities. The availability of non-QoS-capable protocols in providing basic connectivity is also considered a mandatory requirement. The RIP has been selected for this purpose. With regard to QoS support in the network devices, we impose the availability of different queuing disciplines and congestion-avoidance mechanisms. Among queuing, we include weighted fair queuing (WFQ), committed access rate (CAR) control, weighted round robin (WRR), class-based queueing (CBQ), priority, and rate limit. The RED is selected to be the congestion-avoidance mechanism. Finally, no specific security requirements have been identified.

- *MPLS capabilities.* IP routers constituting the part of the testbed emulating a network provider environment must be MPLS enabled [1]. This includes support for TE facilities such as constraint-based route computation and static/dynamic bandwidth reservation and alternate route provisioning. LDP, reservation protocol (RSVP), and BGP4 with VPN enhancements are considered mandatory signaling protocols. The last one could be useful for the implementation of VPNs for testing purposes.

- *Management interface requirements.* Routers must exhibit management interfaces towards the EMSs based on CLI and SNMP, and it is also desirable that they incorporate COPS. Nevertheless, if COPS were not available, the management policies would be converted to CLI or SNMP commands to be enforced in the corresponding devices. At least the availability of the MIB-II with the MPLS objects group active is required, and the availability of the MPLS-LSR-MIB and the MPLS-TE-MIB is desirable [2].

10.2.4 Requirements on the WDM Infrastructure

The testbed infrastructure is constituted by network elements such as OXCs and OADMs and other devices involved in transmission, regeneration, and switching of optical signals/optical channels. We also consider in the WDM testbed the measurement equipment intended to monitor the network performance parameters, providing forward signaling to the management systems. This includes fault detection and backward signaling from the management systems, including network elements reconfiguration.

As a general requirement for the testbed infrastructure, we assume that the testbed shall be able to emulate a WDM network with a physical topology of at least three nodes [3, 4]. These nodes will be either OXCs or OADMs.

10.2.4.1 Configuration Management Requirements

The testbed will have programmable optical nodes (e.g., a programmable OADM or OXC) interconnected with each other and with optical WDM connections to the IP routers. These nodes must exhibit the appropriate functionality for setting up and releasing lightpaths from the EMS northbound interface. Regular updates of the information regarding the network topology, bank of free wavelengths on every link, and routing tables must also be part of the configuration management capability of the WDM nodes [5, 6].

10.2.4.2 Performance and Fault Management Requirements

The performance monitoring is based on the inspection of a set of physical parameters responsible for the network performance. The testbed will allow the measurement of at least the following physical parameters: optical power, frequency, carrier-to-noise ratio, and BER. These parameters shall be regularly measured for every optical signal and every optical channel. The values of the parameters are directed to the EMS, which shall then signal to the NMS.

Threshold values should be assigned to all of the performance parameters. Two threshold levels should be considered: degradation level and failure level.

The testbed should detect both faults of the failure level and faults of the degradation level and must report them to the management systems.

The following procedure should be carried out if one of the thresholds is exceeded:

- Performance failure (PF) or performance degradation (PD) signals are initiated by a network element, depending on which of the thresholds are exceeded.

- The network element that detects a defect generates special alarms and sends them to the EMS. Table 10.1 presents a set of alarms that should be implemented.

- According to the alarm received, the network element or the EMS invokes protection/restoration algorithms to correct the fault and send the necessary commands backwards to the network elements.

In order to test the emission of these alarms, the testbed should have the facilities to both degrade the optical signal on a link in a controllable manner and monitor the optical signal on that link. Also, the testbed should have the facility to interrupt the lightpath in a controllable manner.

10.3   The WINMAN Testbed

10.3.1   The IP Layer

An IP testbed fulfilling all the mandatory requirements mentioned here is depicted in Figure 10.2 and consists of three routers, PE-1, PE-2, and P-3, linking three LAN sites. Also represented in the figure is the EMS responsible to manage the routers and the WINMAN platform under test.

The routers are CISCO 7200 and are intended to adopt the roles of IP network provider devices. Therefore, the first two are provider edge (PE) routers and the third one is a provider (P) router emulating the core network. Each of the three routers has three fast Ethernet interfaces. Two such interfaces are needed to establish the triangle-shaped network of Figure 10.2. The third Fast Ethernet port is used to connect traffic analyzer equipment with fast Ethernet interfaces to the routers, although is not shown in the picture. Each router also

Table 10.1

Alarms That Must Be Implemented in the WDM Testbed Infrastructure

Faults Detected by the Transmission Layer	Faults Detected At the Optical Path Level
Loss of optical signal (LOS)	Lightpath Blocking (LPB)
Loss of payload (LOP)	Lightpath misconnection (LPM)
Loss of optical overhead (LOH)	
Loss of a single WDM channel (LOC)	
Shift of an optical channel frequency (SOF)	
Optical channel degrade (OCD)	

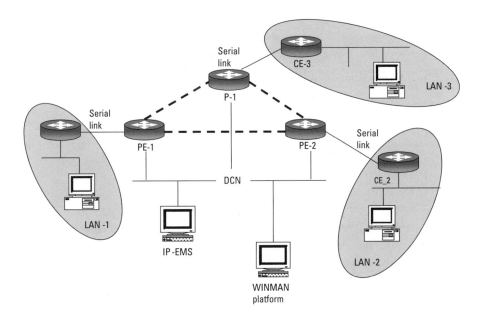

Figure 10.2 The IP layer testbed.

has serial links used to connect to the customer LANs as well as to the data com-munication network (DCN). The Ethernet ports are dual 10/100 ports, operat-ing either at 100 Mbps (100BaseTX) or at 10 Mbps (10BaseT).

Note that the links represented in Figure 10.2 with a dashed line are not physical links. In fact, the PE and the physical routers are connected through the WDM testbed equipment described later in the chapter.

10.3.2 The WDM Layer

The WDM testbed is depicted in Figure 10.3. It consists of a hub and three remote nodes, connected via a two-fiber counter-rotating ring. On the outer fiber ring, dedicated (fixed) wavelengths $(\lambda_1, \lambda_2, \lambda_3)$ are used for communication between the hub and the remote nodes. On the inner fiber ring, each node is provided with a programmable OADM to set up wavelength paths with the other remote nodes or with the hub. As will be described in Section 10.3.3, every node is equipped with a GbE switch for interconnection between the IP routers and WDM nodes. At the hub, IP traffic can be also be routed to a loca-tion outside the testbed (e.g., the Internet). The remote nodes are located at business sites where IP over 10/100 Base-T is the major service. Therefore, the transport stack IP/GbE/WDM at 1.25 Gbps will be used. The GbE switches in

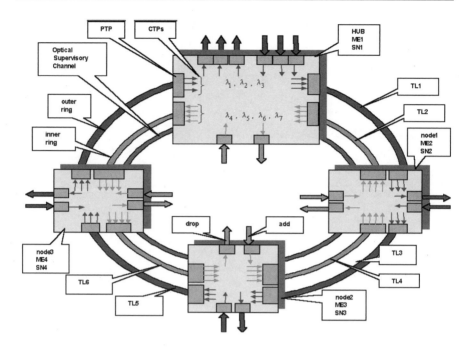

Figure 10.3 The WDM layer testbed.

the testbed are equipped with WDM lasers, which allow them to interface directly with WDM layer.

The distance between two remote nodes is 8 km on average, with a total ring circumference of 32 km. An optical supervisory channel at 1,310 nm will be used to transport management information.

The testbed will contain the following management systems:

- *A network element agent per optical node.* This software presents a management view of the OADMs to upper level management systems that control the optical components using an RS-232 link.

- *An EMS, which will control the four agents.* The information model and the northbound interface of this system have been implemented according to the TMF-MTNM standards.

Figure 10.3 represents the connectivity details of the WDM testbed, which includes four programmable nodes, eight topological links, 12 PTPs in the hub, eight PTPs in each node, and the number of CTPs per PTP varies between one CTP (add/drop) to four CTPs.

10.3.3 Interconnection of the IP and WDM Nodes

The IP/MPLS router can be equipped with different optical interfaces transmitting in a specific wavelength. The OADM could be equipped with tributary cards with or without the capability of adapting the wavelength in the specific grid necessary for the multiplexed signal to have the appropriate spacing. The equipment intended to implement this adaptation is a transponder that executes the task through an optical/electrical/optical adaptation.

The transponders operate in the following way. Each transponder has four interfaces: two interfaces, one of which is a *transmit* and the other a *receive*, towards the client router and another two transmit-receive towards the inside of the WDM system. The receiver from the client router is usually wideband, while the transmitter towards the client is either in the 13xx nm or in the 15xx nm window, according to the distances needed to reach (in 15xx, the distances are greater because the attenuation is smaller). On the other side, the interfaces towards the internal of the WDM are the specific wavelengths added (transmission) to the multiplexed signal or dropped (reception) from the demultiplexed signal. In case of tuneable lasers in the path towards the internal of the WDM, the transmitter wavelength could be variable and adapted to the specific lambda needed.

Because transponders, which are quite expensive elements, were not available in the testbed, alternative approaches were examined. The solution finally adopted is shown in Figure 10.4. Specifically, the wavelengths used for the interconnection need to be colored; this means in specific wavelengths according to the grid used by the WDM ring. In other words, colored interfaces do not need adaptation and can be used directly with the WDM equipment. The Avaya Cajun switches that are part of the WDM testbed provide GbE interfaces with

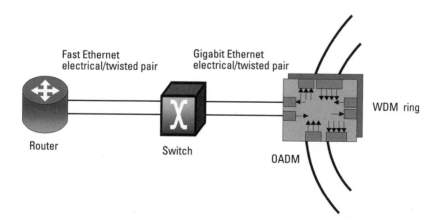

Figure 10.4 Interconnection between the IP layer and the WDM layer.

gigabit interface converters (GBICs) operating at specific wavelengths. In this sense, they are already adapted to the wavelength needed, and they can be directly added to the multiplexed signal. In addition, the wavelength that is dropped towards the switch is directly passed to client receiving interface, which as mentioned before is wideband.

10.4   Adaptation of the Technology-Dependent EMSs to the WINMAN Southbound Interface

Once a testbed has been configured, the NMS under test has to be connected to it through a DCN. For WINMAN, the testbed represents a setup that includes a fully functional WDM network supporting the transmission of IP services over its physical layer. This setup is based on network elements that may have proprietary management interface (i.e., interface towards the DCN). Consequently, EMSs responsible for those elements are equipped with their native APIs implemented in a proprietary way. On the other hand, although the EMSs might provide the functionality specified for the testbed, they might not support the interface with the NMS under test. Therefore, an adaptation layer between native testbed platforms and the WINMAN southbound interface is to be introduced. Although the context of this section is WINMAN oriented, the methodology can be easily applied to any other NMS solution. This methodology consists of the specification and design of an adaptation component on top of each EMS (IP and WDM). These adaptation components appear as a top layer of the testbed infrastructure, and they are the only testbed entities visible from WINMAN. While those components are proprietary for each testbed, they will provide northbound interfaces compliant to WINMAN southbound interface specifications. Figure 10.5 presents the adaptation architecture.

10.4.1   Adaptation of the WDM EMS

Adaptation of the WDM-EMS is based on international standards and, especially, on the TMF-MTNM interface for the WDM-EMS to NMS interconnection. The TMF-MTNM provides the interface between the NMS and EMS, where NMS refers to any client and EMS refers to any server for the interface.

The structure and composition of the adaptation is described by its information model. This model gives the objects that will be presented to NMS. Instances of these objects are contained in the database of the EMS. This database has a static and a dynamic part. In the static part, the physical configuration is represented, like the network elements, their physical ports, and the links between those ports. The dynamic configuration contains the connections that are present in the network elements, together with their CTPs. If a new optical

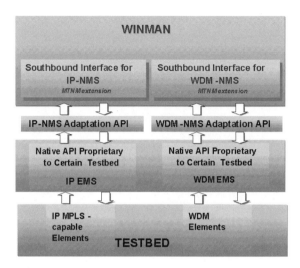

Figure 10.5 WINMAN testbed adaptation architecture.

network element is added, the static configuration changes; if a new subnetwork connection is created, the dynamic configuration is altered.

The WDM software whose architecture is shown in Figure 10.6 was developed on Windows NT using Visual C++. The resulting executables are intended for Windows NT computers. The CORBA version is TAO/ACE by Douglas C. Schmidt (ACE version 5.1.3, TAO version 1.1.3). The WDM-EMS consists of an executable and an initialization file. The initialization file is read at the startup of the EMS. The initialization file contains information that is known only by the operator of the network (e.g., topological links, subnetworks, EMS name, and reporting mode for debugging).

10.4.2 Adaptation of the IP EMS

The adaptation between the WINMAN management system and the EMSs is also based on the TMF MTNM specification. The IP-EMS is a CORBA component that communicates using this interface. It is the responsible entity for the direct interaction of the management system with the devices. Due to the lack of preexisting EMSs, the IP-EMS in the WINMAN testbed is defined more as a mediation device between the NMS and the network elements than a complete EMS. The IP-NMS to IP-EMS adaptation will ensure the establishment of LSPs with bandwidth constraints. The IP-EMS will perform this basic functionality, provided that the network basic IP/MPLS configuration is already set up. The configuration of the IP routers is performed by means of CLI with additional SNMP support for monitoring purposes. The interface with the

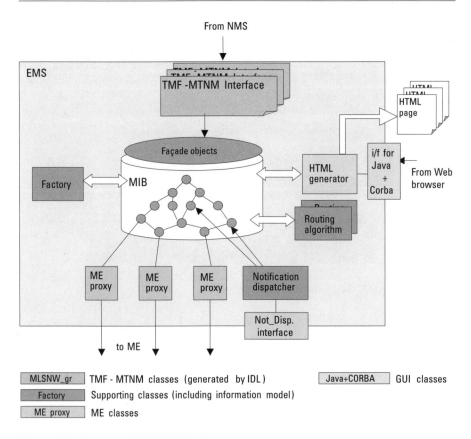

Figure 10.6 Structure of WDM-EMS adaptation.

routers has been developed according to an API whose example is provided in Section 3.3.

10.5 Summary

The deployment of an IP over WDM network testbed is instrumental for the validation of the WINMAN integrated management system. Initially, a set of requirements is classified, looking at the management equipment, the testing and verification tools, and the involved transport network infrastructures.

The general characteristics of the management platforms to be supplied by the testbed are divided into high-level requirements and functional require-ments. High-level requirements are intended to make possible the basic interac-tion between the NMSs and the managed network. One of these requirements is, for instance, referring to the interconnectability between the testbed and the WINMAN system. On the other hand, the functional requirements refer to the functions that the testbed, and particularly the EMSs, have to exhibit in order to check the management functionality of the NMS prototypes in the three management functional areas for which it has been conceived (i.e., to allow configuration, fault, and performance management).

The IP layer of the testbed consists of three MPLS-enabled routers linking three LAN sites. Also belonging to the IP part is the EMS responsible to manage the earlier mentioned routers and the traffic analyzers. Likewise, the WDM testbed consists of a hub and three remote nodes, connected via a two-fiber, counter-rotating ring. On the outer fiber ring, three dedicated wavelengths are used for communication between the hub and the remote nodes, while on the inner fiber ring, each node is provided with a programmable OADM to set up wavelength paths with the other remote nodes or with the hub. Moreover, the WDM part also contains a network element agent per optical node that is connected via RS-232 to the EMS controlling the four agents. Cajun switches that are also part of the WDM testbed provide GbE interfaces. These devices are used to interconnect the routers to the optical nodes because they have GBICs operating at specific wavelengths, thus becoming adapted to the wavelengths needed in the ring.

Although the available EMSs provide the management functionality specified for the testbed, they do not support in principle the interface with the NMSs under test. Therefore, an adaptation layer between native testbed platforms and the WINMAN southbound interface is to be introduced. Adaptation of both the WDM-EMS and the IP-EMS is based on international standards and, especially, on the TMF-MTNM interface for EMS to NMS interconnection.

References

[1]   Awduche, D., et al. "Multi-Protocol Lambda Switching: Combining MPLS Traffic Engineering Control With Optical Cross-connects," draft-awduche-mpls-te-optical-04.txt, April 2002.

[2]   Nadeau, T., et al., "Multiprotocol Label Switching (MPLS) Management Overview," draft-ietf-mpls-mgmt-overview-02.txt, work in progress, expires December 2002.

[3]   Anderson, J., et al., "Protocols and Architectures for Optical Networking," *Bell Labs Technical Journal*, January–March 1999, pp. 105–123.

[4]   Chlamtac, I., A. Ganz, and G. Karmi, "Purely Optical Networks for Terabit Communication," *IEEE INFOCOM*, 1989.

[5]   Strand, J., A. L. Chiu, and R. Tkach, "Issues For Routing In The Optical Layer," *IEEE Communications Magazine*, February 2001, pp. 81–87.

[6]   "High Bandwidth Optical WAN's," *IEEE Transactions on Communications*, July 1992, pp. 1171–1181.

11

Evaluation of IP over WDM Management Solutions

11.1 Introduction

The management of IP over WDM networks has been addressed in various chapters in this book. Moreover, the WINMAN solution has been specified and implemented as a feasible approach for deploying IP over WDM networks. Therefore, it was necessary to verify that this solution was effectively bringing what was intended. And this is the scope of this chapter—to present the experiments that were conducted to assess the WINMAN solution.

The evaluation of a network management solution requires appropriate experiments defined at different levels, covering all of the relevant characteristics. This is not only the system functional behavior but also its nonfunctional attributes. Hence, the testing of an NMS for integrated management of IP over WDM requires a considerable preparation effort due to the complexity of the functional and nonfunctional behaviors to be tested. On the other hand, these tests must be neutral with respect to the target system in order to ensure the reliability of the results. Therefore, the followed approach was to design sets of tests that were generic enough that they were later instantiated for the target system. In this chapter, WINMAN is the target system of the experiments. Nevertheless, the evaluation criteria and experiments were designed to be applicable to any other NMS in the IP and WDM network technologies.

The chapter is structured in six sections. After this introduction, we start by portraying the template to fill in for specifying any type of validation experiment. Section 11.3 is entirely devoted to the nonfunctional system attributes—openness, flexibility, modularity, scalability, portability, and robustness.

We give the meaning of each of them and also outline their testing scenarios. Section 11.4 specifies experiment scenarios for evaluating the functionality of NMSs for management of IP over WDM. The functional attributes to be tested lie in the areas of configuration, fault, and performance management and are further decomposed according to the TMF TOM [1] processes areas as follows:

- Network provisioning and network inventory for configuration management;
- Network maintenance and restoration for fault management;
- Network data management for performance management.

Only the first of the group of three is further worked in the book. Based on the specified experiments, Section 11.4 evaluates the WINMAN system under conditions close to a real network operational environment. The experiments were carried out by means of a network setup, where the WINMAN system was integrated with an IP over WDM testbed, as described in Chapter 10.

11.2   Template for the Definition of Testing Experiments

All of the experiments carried out for the validation of WINMAN attributes have been defined by filling the description template. The fields and meaning of this template are summarized in Table 11.1.

11.3   Evaluating the Nonfunctional Behavior

Like any software application, an NMS is assumed to exhibit behavior that does not pertain directly to its dedicated functionality, yet it furnishes software with high quality. Amongst the most critical quality characteristics are the nonfunctional attributes: openness, portability, flexibility, modularity, scalability, and robustness. The meaning of these attributes and the scenarios that were envisaged to test them in the context of the book follows hereafter.

11.3.1   Openness and Flexibility

Openness refers to the interlocking of the system with other components or systems, likely residing in different locations. Portability is the system's ability to be executed in different processing nodes without modification.

A scenario intended to evaluate the WINMAN NMS system's openness and portability characteristics was conceived. Its particular objectives were to demonstrate the interworking between components residing in different systems

Table 11.1

The Experiments Template

Field	Description
Priority	The priority of the experiment is qualified as *high, medium,* and *low,* according to the relevance to the scope, the feasibility, and the costs of the related experiment.
Objectives	Describes the aim of the experiment. The description must indicate the property of the system under evaluation, referencing the system requirement where this property is specified; the significance of the experiment; the reasoning behind it; and the scope and boundaries.
Background	Provides information related to the property of the system that will be evaluated. This might contain theoretical background, references to standards, or a business model.
Required tools	Management platforms and auxiliary equipment. It refers to any kind of hardware and software complementing the management platforms that may be needed for executing the experiment. This includes, for instance, measurement equipment like traffic analyzers, optical analyzers, and end-user applications.
Required infrastructure	Description of the testbed transport network infrastructure, including possible interconnectivity between different testbeds, the data communication network, and the connectivity to the end users. If a network emulator is required, it should be specified here.
Expected outcome	Describes the expected results from the experiment. The expectation may be expressed with respect to standards, requirements, or service performance figures.
Experiment setup	Describes the setup of the experiment, which may include the transport network infrastructure configuration, the management system configuration, points of measurements, and preprovisioning of the management system and the managed network if necessary.
Scenario	Describes the steps to be executed and any precondition and postconditions that must be ensured.
Related experiments	Provides references to related experiments.

and that the NMSs were able to open up the network resources, as standard interfaces, for use by various network management applications. This capability was expected to be tested by making use of a MIB/IDL browser that should be able to browse through the MIB objects. Another specific objective was to see that the NMSs were able to operate in various operating systems, including at least Solaris and Windows NT 4 or Windows 2000. The functionality should remain the same regardless of the operating system used.

The expected outcome of the experiment is that the NMS will operate on all platforms without affecting its functionality and will interface with other network management platforms.

11.3.2 Flexibility and Modularity

Flexibility means being capable of both evolving and accommodating the existence and continued operation of legacy or new systems. A flexible distributed system should be capable of facing run-time changes; for example, it should be capable of being dynamically reconfigured to accommodate changing circumstances. On the other hand, a system is said to be modular if it consists of modules (software modules) that can be removed, replaced, and upgraded without affecting other modules that are already in place. This avoids the creation of monolithic applications that make modifications to the software cumbersome.

A scenario was designed to evaluate the NMS system's flexibility (in terms of accessibility, new applications integration, different levels of functionality, and network elements scope) and modularity (in terms of ease of module expansion and module upgrading). The desired outcomes of the experiment are that the NMSs will operate normally when different management applications are started or shut down from different workstations. The software modules that comprise the NMS will be able to be stopped, removed, replaced, recompiled, and reintegrated to the system. New applications/modules that are introduced can use the existing management information bases and the historical data.

11.3.3 Scalability

Scalability refers to the impact on *performance* as more entities (management applications, processes, and network nodes) are added to the system. Performance is a synonymous term of system throughput. This is often traded off against functionality because the more a system tries to do, the lower its throughput is. Any degradation of system performance can, if magnified, significantly affect a system's overall integrity. The way a system scales is to a great extend a function of its communication complexity, computational complexity, data complexity, and time complexity. The communication complexity gives an assessment of either how many operations or messages need to be generated to perform some action. The time complexity expresses how long (in cycles or iterations) different operations take. Computational complexity relates to how well a given procedure can be analytically described or determined. Data complexity refers to the complexity of data structures and their interdependencies.

The specific objective of the scenario designed to test the WINMAN solution scalability is to validate that an increase in the number of network elements will not cause major performance problems to the NMSs. It is expected that the system will still perform without interruption with a great number of managed objects and without major performance problems. A specific prerequisite for the *infrastructure used in the experiment* is the capability to emulate additional network elements so that we can work with networks containing about thousand WDM and IP nodes.

11.3.4 Robustness

Robustness can be understood as the ability of the system to handle unexpected events. It is also related to the system integrity; the system maintains its correct and proper functional profile. The more robust the system is, the more likely it is to retain a high level of integrity within its operational environment. A system that is robust can cope with all eventualities and continue operation (i.e., it does not halt except when required). The opposite of robust is brittle. A system is brittle if it is likely to fail in most circumstances of the operational environment.

Evaluating robustness is a difficult task because it deals with unforeseen inputs and extreme situations. It becomes even more difficult for distributed applications because their concurrent nature causes an interleaving of events that can be difficult to predict. Although the time is the best evaluator of robustness, robustness can be evaluated by fitting the system with extreme data (either extremely large or small data sets or boundary values), bogus data (too large or small or invalid values), and by submitting it to considerable stress.

11.4 Evaluation of Functional Behavior

One of the most important objectives of the WINMAN NMSs testing processes is to show that the WINMAN solution is able to provide and manage ICSs according to the functionality appointed in Chapter 6 of the book. It is assumed that the WINMAN functionality lies in the areas of configuration, fault, and performance management, which will be separated according to the TMF TOM [1] areas as follows:

- Network provisioning and network inventory for configuration management;
- Network maintenance and restoration for fault management;
- Network data management for performance management.

Nevertheless, according to the scope of the book, the experiments proposed aim only to exhibit and validate the network provisioning and network inventory management functionality. Although the main goal is to evaluate the intertechnology functionality of WINMAN, single-domain functionality dealing with IP only or WDM only is also accommodated. This is useful for NMS operators, giving them the opportunity to realize useful or necessary single-domain functionality. As far as single-domain network provisioning functionality, the creation of IP/MPLS and optical paths scenarios will be described for the IP and WDM domains, respectively. The network inventory will be focused on the correct depiction, through the GUI, of the managed network infrastructure.

It is worth mentioning that the management solution under test is policy based. Under this assumption, every experiment performed on the system will lead to different results aligned with the current active policies, because policies dynamically change the system behavior. For example, the *ICS provisioning* experiment can have different outputs if some policies are changed from one round of the experiment to the next.

11.4.1 Configuration Management/Network Provisioning/Create Optical Path

It is essential for the WINMAN operator to be able to manually provide end-to-end optical paths, either during the initialization and configuration of the IP over WDM network, following the network planning period, or for any other reason (e.g., fault, performance degradation, or capacity exhaustion) that the NMS was unable to automatically resolve. In addition, controlling WDM network resources states and their modifications is a basic management action for the NMS configuration functionality.

The objective of this experiment is to evaluate and demonstrate the creation of an end-to-end optical path in the WDM network by the system operator and the capability of the WDM-NMS for communicating with the corresponding EMS of the testbed. This experiment scenario is related to the scenarios that show the GUI functionality and is part of the scenario showing interdomain connectivity functionality.

11.4.2 Configuration Management/Network Provisioning/Create IP/MPLS Path

The experiment is to verify that the IP-NMS is able to provide all of the consecutive procedures for the provision of IP connectivity with guaranteed bandwidth and QoS by means of a MPLS LSP.

A testbed or service provider owner of IP equipment can only use this experiment in order to test the basic IP-NMS functionality of creating end-to-end connections over the IP/MPLS infrastructure. From this point of view, it makes sense to have this pure IP/MPLS experiment separated from the general ICS provisioning, which involves a complete IP/WDM network and management infrastructure.

This experiment scenario is related to the scenarios that show the GUI functionality and is part of the scenario showing interdomain connectivity functionality.

11.4.3 Configuration Management/Network Provisioning/ICS Provisioning

Different end-user services (MoIP, VoIP, and VPN) will require different connectivity services from the network. This experiment will prove that the NMS solution is able to accept and process a VPN service. Table 11.2 shows the specific aspects of this experiment.

Table 11.2
Experiment Template for Testing the ICS Provisioning

Field	Value
Priority	High
Objectives	The NMS serves a request from an SMS for the provisioning of IP network connectivity that supports the VPN service to the end users. The NMS system will show its flow-through provisioning capability (i.e., that the request will be processed fully automatically from request reception to service operational). The operability of the service will be made visible by means of end-user applications making use of the network connections. The data stored in the NMS databases will be checked for validity and consistency. The flow-through provisioning capability is one of the main features of the solution, which makes this experiment very significant.
Background	A company having several sites spread over the country wants to interconnect the LANs of each site via an IP network with high-speed connections, a high security level, and determined QoS. The solution will be an IP-based VPN. The VPN must support real-time and nonreal-time services. The SMS system orders the network connectivity over an interface, which is a publicly specified interface.
Required tools	End-user applications that generate traffic that will flow over the provisioned network connections and traffic analyzers that can measure the QoS parameters of the user flows.
Required infrastructure	An IP over WDM network as described in Chapter 10.
Expected outcome	The IP connectivity is delivered through the network. The end-user applications are able to make use of the delivered network services. The performance of the network connections will be measured by means of IP traffic analyzers and will be reflected by the quality perception of the end-user applications.

Table 11.2 (continued)

Field	Value
Experiment setup	End-user equipment is connected to IP edge routers. On top of this equipment, end users will run their applications, like VoIP and MoIP, which generate real-time traffic. Nonreal-time applications, like file transfer, database access, and Internet browsing, may also run on this equipment.
	An SMS is connected to the NMS. The former may be a commercial system, which is able to accept requests from end users, or it may be an SMS simulator, which is preprovisioned to generate a request to the NMS under test.
	The IP and WDM NMSs (this is the technology-dependent management systems) are connected to IP and WDM EMSs in the testbeds via the southbound interface.
	Test equipment is connected to certain measurement points in the network for the monitoring of the provisioned connections.
	Auxiliary windows are available on the GUI for the monitoring of the contents of the databases.
Scenario	This scenario could be executed with different service requests having different values for the service attributes.
	Preconditions:
	The managed network is operational and the basic network configuration is provisioned.
	The NMS is operational, the GUI is operational, and the databases contain all relevant data of the network.
	The SMS system, the end-user equipment, and applications are operational.
	Scenario steps:
	The end user sends a request for VPN connectivity to the SMS. The SMS sends a request to the NMS under test.
	The NMS processes the requests and executes all of the steps of the flow-through provisioning process without any manual intervention from the operator.
	During the provisioning process, the technology-dependent NMSs communicate over the DCN to the EMSs.
	Monitor the start and performance of the end-user applications.
	Read out the test equipment.
	Read the NMS databases.
	The experiment scenario ends.
Related experiments	Create optical path and create IP/MPLS path.

11.4.4 Network Inventory Management/Network Inventory Consistency to Network Infrastructure

This scenario will evaluate the network inventory functionality of the NMS system. This is the NMS capability to maintain the network inventory in an accurate and consistent manner with respect to the underlying network infrastructure. The NMS system, upon initialization, should not only discover correctly the underlining network infrastructure and store it to the network inventory; it should continue to check its consistency (i.e., modification, removal, or addition of links or nodes). Through the GUI, the NMS operator can evaluate the network inventory stored information.

The presentation of a precise and consistent view of the managed network is considered as a basic functionality for an NMS and is part of all other experiments.

11.5 The WINMAN Configuration Management Solution Evaluation

In order to evaluate the WINMAN system under the conditions close to a real network operational environment, the experiments were carried out by means of the network testbed described in Chapter 10. Besides the IP/WDM testbed and the WINSMAN system, application layer devices are accommodated as depicted in Figure 11.1 in order to get rid of the WINMAN functionality in the context of a business case.

11.5.1 Integrated Experiment for Validating the WINMAN Functionality

The WINMAN functional attributes will be validated using an integrated network provisioning scenario. The main purpose of this scenario, as shown in Figure 11.1, is to establish an ICS between routers PE-1 and PE-2 using the P-3 intermediate node, instead of the shortest path, due to QoS constraints. The IP link between PE-2 and P-3 is configured but not operational because there is no physical (WDM) connectivity yet. In order to provide an ICS, the management system has to first establish the physical link between PE-2 and P-3 and then create a bidirectional LSP between the PE routers through the provider router. The business case behind this scenario is that the WINMAN system is able to find the resources and provide an ICS with QoS constraints, in a case where the default shortest path could not do so.

This scenario involves multiple provisioning requests, and all of the scenario steps along with the start-up conditions are listed next.

11.5.1.1 Start-Up Conditions

There are several start-up conditions.

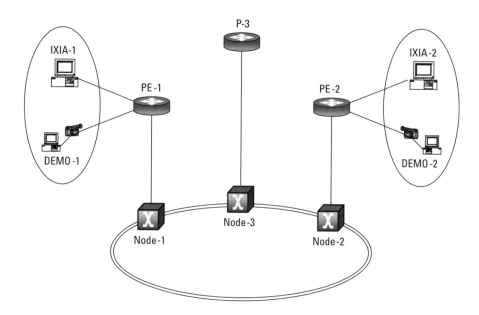

Figure 11.1 Physical entities used in the experiment.

1. There are two LAN sites as shown in Figure 11.1:

   • LAN-1 contains the multimedia customer Demo-1 and a traffic generator IXIA-1 intended to create the interfering traffic in the site.

   • LAN-2 contains the multimedia customer Demo-2 and a traffic generator IXIA-2 intended to create the interfering traffic in the site.

2. There is pure layer-3 (IP) connectivity between routers PE-1–PE-3 and between routers PE-1–PE-2. The IP link between PE-2 and P-3 is configured but not operational because there is no physical (WDM) connectivity, so communication between PE-2 and P-3 is feasible only through PE-1.

3. The 10/100 Ethernet ports between the core routers are configured at 10 Mbps so that we can easily saturate the direct link PE-1–PE-2. This kind of configuration also limits the bandwidth available in each WDM λ to 10Mbps. This does not decrease the functionality of the network because it will only decrease its bandwidth to values that are more suitable for testing purposes.

4. The WINMAN system has the following policies activated:

   • *Authorization policy for the interference traffic.* In order to establish an IP link for interference traffic between PE-1 and PE-2, there must be a policy that specifies the authorization to do it. Therefore, a positive authorization policy to create IP SNCs between the involved PTPs will be active.

- *Denying ICS establishment by policies.* Once the interference traffic is established and the poor video quality over the congested link is demonstrated, the WINMAN operator will try to establish an ICS for the multimedia traffic. Here, we will apply two policies: a negative authorization one to avoid establishing a new wavelength path in the WDM layer between nodes 2 and 3 and a second negative authorization policy to disable an alternate IP link for the intended ICS. This is:

  1. Negative authorization between CTPs of WDM nodes 2 and 3;

  2. Disable IP links between PE-1 and P-3.

11.5.1.2 Scenario Steps

This list describes a global scenario decomposed into subscenarios that explicitly show the different functional attributes of the WINMAN solution.

1. Without congestion, the network establishes a videoconference between Demo-1 and Demo-2 using the conventional best-effort IP routing and forwarding. In this condition, good quality of videoconference shall be observed. Each party sends 1.5 Mbps of video/voice traffic.

2. The IXIA-1 customer requests an ICS between PE-1–PE-2 for 9.5-Mbps guaranteed bandwidth. This involves the creation of a bidirectional LSP between both routers with this bandwidth constraint.

3. The WINMAN operator sends a request to WINMAN for this ICS creation.

4. The first policy (authorization for interference traffic) allows the creation of that ICS, so WINMAN proceeds to establish the LSPs between the involved routers, making the necessary bandwidth reservation.

5. The two interference generators start injecting traffic, making use of all of the requested bandwidth.

6. In this network condition, the quality of the best-effort videoconference between Demo_1 and Demo_2 will be affected. This substantiates that the WINMAN-reserved MPLS traffic is prioritized over the best-effort video traffic that is partially dropped, resulting in a non-pure video reception quality.

7. The WINMAN customer DEMO wants to upgrade its service from best-effort to "gold," requesting a bandwidth reservation of 1.5 Mbps between its two clients DEMO-1 and DEMO-2.

8. The WO requests to WINMAN the new ICS.

9. Because there is not enough capacity between the two (from the 10 Mbps available, 9.5 Mbps are already reserved for the IXIA customers), the WINMAN system decides that a new optical path should be created between nodes 2 and 3 in order to create an IP link between the two that would enable the creation of the bidirectional LSPs PE-2–P3–PE-1 and PE-1–P3–PE-2.

10. The creation of the new optical path cannot be accomplished because the second policy (negative authorization between CTPs in WDM) does not allow it. The WINMAN operator gets a failure message from the system indicating that the rejection is due to applicable policies. As a consequence, the WINMAN operator checks the current policies, changes the specific policy from negative to positive authorization, and triggers the ICS creation again.

11. The WDM optical path (node-2 and node-3) is created. The creation of the new optical path results in the autodiscovery of a new IP link. Hence, the IP-EMS detects the new link and sends a notification through the corresponding WINMAN southbound interface that finally arrives to the WINMAN INMS.

12. The IP-NMS carries out a checking process of the new link.

13. The WINMAN system tries now to establish the ICS though the router P-3. Nevertheless, the creation of the bidirectional connectivity PE-2/P-3/PE-1 and PE-1/P-3/PE-2 is not possible because of the prevailing negative authorization routing policy.

14. Again, the WINMAN operator checks the active policies and changes the negative authorization policy for a positive one.

15. The ICS creation for the DEMO client is retriggered by the WINMAN operator. Now the request gets a successful result.

16. The quality of the videoconference between the DEMO client sites turns to good even though the IXIA client is still injecting the same amount of traffic. This concludes the experiment.

11.5.1.3 Network Provisioning/Create Optical Path

The creation of a new optical path is needed to support the ICS requested by the DEMO customer.

The WINMAN system is requested to create an ICS to serve the traffic flow generated by the multimedia clients. The requested bandwidth is

1.5 Mbps. The WINMAN system will find the shortest path between PE-1 and PE-2; when it checks in the inventory, the available resources it will find that there are not enough for the request. Then, the WINMAN system will look at the WDM network and reach the conclusion that there is the possibility to create an extra optical trail connecting PE-2 to P-3. The request for the creation of the optical trail is forwarded by the WINMAN system to the WDM-EMS.

11.5.1.4 Network Provisioning/Create IP/MPLS Path

The IP/MPLS path supports the ICS in the WINMAN system. In the integrated scenario two bidirectional ICSs are created. These ICSs are supported by bidirectional SNCs at the IP level, which involves the creation of two IP/MPLS unidirectional paths for each SNC. Therefore, the total number of IP/MPLS path created in the integrated scenario is four.

11.5.1.5 Network Provisioning/ICS Provisioning

As stated earlier, two bidirectional ICS are provided in the integrated scenario:

- *The ICS for the IXIA customer.* The interference generators are WINMAN clients that need to exchange information. So they request for a WINMAN ICS between PE-1 and PE-2, saying that they want to have 9.5 Mbps always available. The WINMAN system creates the ICS, making the necessary bandwidth reservation and establishing the LSP between the routers. The two interference generators start working and they use all the bandwidth they requested to the WINMAN system. The traffic generators used are the Lucent IXIA product [2].

- *The ICS for the DEMO customer.* Once the new optical path is created and the new IP link is detected, the WINMAN system will recognize two IP paths between PE-11 and PE-2. The first one is the shortest path, and the second one is through P-3. The shortest path does not have enough bandwidth and the WDM capacity is exhausted, so the other option is used to accommodate the new ICS. The traffic generated by the multimedia clients is now forwarded through the new ICS, and the video quality becomes the desired.

11.5.1.6 Network Inventory/Network Inventory Consistency to Network Infrastructure

The new optical trail is created and a new IP link is discovered by the IP EMS between PE-2 and P-3. The WINMAN network inventory is updated with the new WDM and IP links, and the GUI is displaying the actual network configuration.

11.5.2　Experiment Results

This section describes WINMAN system reaction observed while executing the experiments for functional and nonfunctional attributes. Evaluation of the experiment results against the required functionality for both functional and nonfunctional attributes is summarized in the tables hereafter.

11.5.2.1　Experiments Validating Functional Attributes

All of the experiments defined in the previous section were completed successfully.

11.5.2.2　Experiments Validating Nonfunctional Attributes

The experiments validating nonfunctional attributes were executed explicitly or implicitly. Modularity, flexibility, portability, and openness are nonfunctional attributes that are inherently supported by the WINMAN solution architecture and the core technologies that were used in order to implement it. The choice of using standard interfaces, like CaSMIM and MTNM, gives the WINMAN system the desired openness. Moreover, the WINMAN solution is implemented using Java programming language, which is highly platform independent. Additionally, the implementation is encapsulated in components that are provided by the component-based platforms that were employed—DSC and NETRAC. DSC uses CORBA middleware as the core communication mechanism between components, which brings along all of the valuable attributes of this technology. We can add components to the WINMAN solution imposing minimal alterations to the existing functional system in a time-efficient way for both development and integration. These new components can support new entities (applications, network nodes, new types of interfaces, and new functionality). These components can also replace malfunctioning components without affecting functionality that directly makes use of them.

More specifically, the results from the experimentation regarding nonfunctional attributes are:

- *Portability.* During execution of the experiments, we installed the DSC platform on a Solaris machine as well as on a Windows NT machine without any side effects on WINMAN overall behavior. Moreover, we managed to run the GUI smoothly from both UNIX and Windows machines.

- *Flexibility and modularity.* During the experiments, we started, stopped, replaced, and recompiled several components (specifically, provisioning manager, routing, and the policy manager) without the need for rebuilding the entire WINMAN system.

- *Robustness and scalability.* We did not stress the system to its limits because it is in the development phase, and some minor communication issues between components are still under investigation. Moreover, the autodiscovery of new network elements was not available, so it was impossible to create many network elements by means of an emulator.
- *Openness.* To test that WINMAN is open to other, higher level management systems, we sent SMS commands using tcl/tk scripts [3]. The results were completely satisfactory.

11.5.2.3   Result Sheets

The results of the experiments are summarized in Tables 11.3 and 11.4.

Table 11.3

Results of the Functional Attributes Validation

Experiment	Status	Result	Comments
Create optical path	Executed	Optical path between node-2 and node-3 was established.	The experiment was executed using the DSC actors. The result was confirmed by observing the status of the specific equipment (the green LEDs were on) and the GUI.
Create IP/MPLS path	Executed	IP/LSP path was established after removing the negative authorization policy.	The process of setting up the path took up to 15 minutes, with all debugging features activated. The experiment was performed using the DSC actors. The result was confirmed by multimedia traffic.
ICS provisioning	Executed	ICS was created.	All of the correct steps involved in ICS creation were taken. However, they were performed as separate actions and not as one go. The initiation was performed through the WINMAN northbound interface.
Network inventory consistency to its network infrastructure	Executed	Inventory was notified of newly created links and the GUI map was updated accordingly.	

Table 11.4
Results of the Nonfunctional Attributes Validation

Experiment	Status	Result	Comments
Portability	Executed	Passed	
Flexibility and modularity	Executed	Passed	
Robustness and scalability	Not executed	N/A	To be executed when the system is in a more mature state
Openness	Executed	Passed	

11.5.3 Evaluation of the Results

Even for people not familiar with other network managers, the WINMAN system does not need specific introductory aids. Getting acquainted with all of the different views in the GUI is not time consuming, as each view (INMS, IP-NMS, and WDM-NMS) has the same philosophy of displaying termination point, links, and connection services. Moreover, the policy manager user interface is launched directly from the main GUI window for additional convenience. During all experiments, the GUI proved a helpful mate.

The GUI always presents the most up-to-date status of the managed network to the WINMAN operator. There were no misalignments between network status and network map. This validates also the proper functioning of the southbound interface and network inventory manager. End-to-end routing guided by the provisioning manager exhibits sophisticated router-based methods, speeding discovery of available or potential routes.

Some minor remarks are that the nodes and termination points are not easily distinguishable, and in case of many depicted elements the screen should be adjusted to higher resolutions to have a better view of the underlying network. All nodes (PTPs, devices) in the views can be dragged on the screen so that they can be allocated in a manner that is readable to the operator. Moreover, the views are not automatically refreshed when a change takes place in the system. It is only updated with the changes when it interrogates the system using the refresh button. That is why there is a refresh button, by which, when pressed, the GUI retrieves all of the newly available information.

During the execution of experiments, every step was performed using the WINMAN GUI and not by sending requests by an SMS system. Nevertheless, to test that WINMAN is open to other higher layer management systems, we sent SMS commands making use of tcl/tk scripts [3] emulating the whole process, and the outcome was as expected. The integration of WINMAN with an SMS platform is expected to be further pursued.

11.5.3.1   General Remarks

The management traffic was not encrypted and the administrators are recommended to exercise caution when using a public network for remote management. However, in most cases the DCN for linking the management systems is a private network or at least a VPN [4].

The waiting time between the issue of a request for the creation of a connectivity service and the display of the new link in the network map was very long. However, it should be noted that due to further development of WINMAN software, every possible debugging feature was enabled, resulting in additional latencies of the system.

The experimentation demonstrated that a system for integrated configuration management is a feasible and valuable solution for the deployment of IP over WDM networks.

11.5.3.2   Overall Assessment

During the experiments, it was validated that WINMAN exhibits and leverages network relationships. The WINMAN system succeeded in carrying out the provisioning of IP connectivity services with guaranteed QoS (in terms of guaranteed end-to-end bandwidth provisioning) in an automatic and fast way by making the appropriate changes to the IP and to the WDM networks. Having knowledge of both network layers, network resources are exploited in an easy and consistent way under the supervision and flexibility provided by management policies. The adaptability and the integrated features of the provisioning manager of WINMAN are key aspects that are absent from all current commercial products.

In conclusion, experimental scenarios for evaluating and demonstrating integrated management solutions in an IP over WDM network were specified for both functional and nonfunctional attributes. These specified experiments were executed in a combined scenario exhibiting the added value of WINMAN solution first release (configuration management only). Providing an integrated IP and WDM configuration management is only half way towards for an integrated solution for managing IP and WDM networks. Fault and performance management, and especially alarm monitoring and network restoration, in conjunction with SLA and QoS management, are challenges that are being addressed in a new release of the WINMAN solution.

11.6   Summary

In order to evaluate a solution that addresses the management of IP over WDM networks, appropriate evaluation scenarios are designed, which demonstrate the competence and wholeness of the management application. Among the most critical quality characteristics are the nonfunctional attributes: openness, flexibility, modularity, scalability, portability, and robustness. Although nonfunctional attributes are quite important for an NMS, the main substance of an integrated NMS is its functional characteristics. The WINMAN solution was evaluated to see if was capable of providing managed end-to-end ICSs and, more specifically, if it was capable of performing integrated provisioning of IP/MPLS LSPs over optical paths. The experiments that are conducted for the functional *certification* of the WINMAN application are lightpath creation, IP/MPLS path creation, ICS provisioning, and network inventory consistency to network infrastructure.

The WINMAN functional attributes are validated using an integrated network provisioning scenario. The business case relevant to this scenario is that when the available IP resources cannot provide the requested QoS, a new WDM path should be established by discovering additional resources, resulting in an additional IP link. This additional IP link enables the creation of an alternative MPLS LSP to serve the needed QoS.

The WINMAN system succeeded in carrying out the provisioning of connectivity services with guaranteed QoS (in terms of guaranteed end-to-end bandwidth provisioning) in an automatic and fast way by making the appropriate multilayer changes to the underlying network. Furthermore, it was validated by the performed experiments that the WINMAN system, due to the special care taken during the design and implementations phases, inherently supports attributes of modularity, flexibility, portability, and openness, as it can integrate new entities with minimal side effects and can work in a distributed way that avoids or corrects malfunctioning parts. As an overall assessment, the evaluation process demonstrated that a system for integrated configuration management is a feasible and valuable solution for the deployment of IP over WDM networks.

References

[1]   TMF, "Telecom Operations Map," Version 2.0, November 1999.

[2]   The IXIA product, www.ixiacom.com/products/chassis/IXIAChassis.php, December 2002.

[3]   Zeltserman, D., and G. Puoplo, *Building Network Management Tools with Tcl/Tk*, Prentice Hall, 1998.

[4]   Katopodis, H., and A. Lyratzis, "Design of Data Communication Network for TMN: The OTE Case," Network Management Development Group, internal report, Athens, Greece, June 10–16, 1998.

12

Integration of IP over WDM: The LION Project Approach

12.1 Introduction

The IST Project LION is a 3-year project led by Telecom Italia Lab and started in February 2000. The Consortium, partially funded by the European Commission, consists of European and Japanese carriers, U.S. and European manufacturers, and leading European universities working together to develop and validate new optical networking standards. Project members include Agilent Technologies Italy, Cisco Systems International B.V., Interuniversity Microelectronics Centre, National Technical University of Athens, Nippon Telegraph and Telephone Company, Siemens Information and Communication Networks, Sirti S.p.A., T-Systems Nova, The University of Mining and Metallurgy, Telekomunikacja Polska SA, Tellium, and the Universitat Politècnica de Catalunya.

The main goals of the IST Project LION are to study, develop, and experimentally assess an ASON/GMPLS carrying multiple clients (e.g., IP/MPLS and SDH). In particular, an ASON is a transport network capable of supporting both classical leased line connections and two other innovative transport services: soft permanent and switched optical connections. As depicted in Figure 12.1, soft permanent connections are set up from the management system, which uses network-generated signaling and routing protocols to establish connections; on the other hand, switched connections can be directly set up and released by the customer on demand using signaling and routing protocols.

Specifically, these innovative functionalities for the dynamic setup and tear down of optical connection are achieved by introducing the so-called

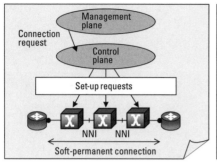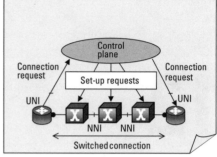

Figure 12.1 Soft permanent and switched connections.

control plane (CP). Basically, through signaling, the CP sets up and releases connections and may restore a connection in case of a failure. ITU-T Recommendation G.8080 [1] describes the set of CP components in terms of abstract entities rather than instances of software components, which are used to manipulate transport network resources in order to provide such functionalities. It should be noted that the principle of automatic switching as recommended in ITU-T Recommendation G.8080 are applicable to SDH transport networks, as defined in ITU-T Recommendation G.803 [2], and to OTNs, as defined in ITU-T Recommendation G.872 [3].

The project was set up by the need of network operators to seek how to both accommodate increasing bandwidth demands for data traffic and provide optical connections rapidly, dynamically, while making optimal use of network resources and saving in operating costs. Therefore, the project addresses the two major drivers for the transport network evolution: optimization of investments and operating expenses and fulfillment of new emerging requirements by transport networks. Specifically, these emerging requirements are reducing the complexity of the transport network (delayering) while optimizing it for the transport of data traffic, introducing intelligence in optical networks for fast and flexible provisioning of high-capacity end-to-end connections, optical rerouting and restoration, interworking functionalities (e.g., for resilience), multidomain interconnections (e.g., network node interface interoperability), traffic engineering for multiple levels of QoS, and provisioning of new services such as optical virtual private networks (OVPNs).

Given the main goal of investigating IP/MPLS over ASON/GMPLS network scenarios, the project activities regard the definition of functional requirements of a next generation transport network, the design and implementation of both optical CP and NNI signaling to enable functionality of set up and tear down of soft-permanent optical connections, the design and implementation of an umbrella management system that enables an end-to-end view over different domains with different management technologies (e.g., SNMP, WBEM, and CORBA), and the definition of integrated resilience strategies for IP/MPLS over ASON/GMPLS. Finally, network planning and evaluation activities will identify convenience areas for IP over OTN and ASON/GMPLS deployments.

12.2 Network Studies

Besides the network demonstrator, an important part of the IST Project LION deals with conceptual and numerical network studies. The diversity in background of the project participants (network operators, suppliers, and academics) has lead to a roadmap that defines the scope of those network studies.

This roadmap is shown in Figure 12.2. The delayering of current data networks is already taking place and should soon result in an IP network directly supported over an OTN. It is expected that within a few years, the OTN will be enhanced with a distributed CP, in order to allow soft-permanent and even switched optical network services in addition to the classical permanent transport network services. While, at first, no integration could take place between the CPs of the IP and the OTN network, the integration of both could open new opportunities in the long term.

If the integration of both CPs does not take place, the network is said to adopt an overlay model. This overlay model is mainly (but not only) driven by the standardization of the ITU-T ASON architecture. If the control of both network layers is integrated in a single CP, the network is said to adopt a peer model. Mainly, the IETF is driving the realization of this CP model by generalizing the MPLS concept.

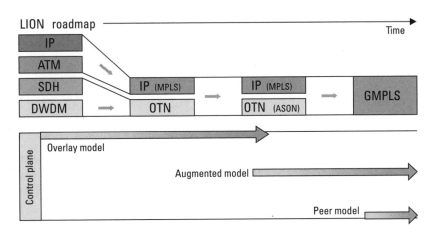

Figure 12.2 Roadmap of the LION network studies.

This section gives an overview of some of the most important ASON-related network studies performed within the LION project. Section 12.2.1 starts by investigating how to model the switched traffic to be carried on an ASON. In Section 12.2.2, some sample areas are depicted in which network investment costs can significantly be reduced by adopting switched optical network services. Finally, some concluding guidelines are presented in Section 12.2.3.

12.2.1 Teletraffic Engineering for Modeling Dynamic Traffic and Dimensioning ASON

Designing an ASON network requires the provisioning of sufficient channels in order to keep the blocking probability below a guaranteed level. Some traffic models (Poisson, Fredericks, and Engset) have been developed to design traditional POTS/ISDN networks. This study investigated whether one or more of these models are suitable to characterize the traffic that is expected to be offered to an ASON network.

The Poisson model is characterized by exponentially distributed interarrival times (IATs) of connection requests and exponentially distributed holding times (HTs). It requires an infinite number of sources, each one offering an infinitesimal amount of arrival intensity. The congestion probability in the Poisson model is given by the well-known Erlang B formula. The Fredericks model (modified Poisson) was developed to catch the characteristics of the overflow in the telephone traffic, introducing an alternate routing strategy. The traffic tries the first routing choice; if rejected, it is offered the next choice. The Engset model assumes a finite number of sources, each requiring one connection at a time. A source may be idle or active (when it has a connection in place); the sojourn time in both states is exponentially distributed. When a source requires a connection, it switches from idle to active state if it finds a free resource; otherwise, it falls in the idle state again.

To assess the applicability of these classical traffic models on ASON networks, their blocking probability performance is compared to the blocking probability of a realistic traffic pattern that could be offered to an ASON. This traffic pattern is obtained through simulation, where an IP router, connected to an OXC in the ASON, generates self-similar traffic and requests to set up or tear down a channel.

As can be seen from the left side of Figure 12.3, the Engset and Fredericks model experience only a slightly higher blocking probability, compared to the simulated ASON traffic pattern, while the Poisson model experiences a significantly higher blocking probability. This can probably be explained by the fact that both the Engset and Fredericks models result in a smoothed traffic pattern (i.e., the variance of the offered traffic is significantly smaller than its mean) similar to the simulated ASON traffic pattern. However, the Poisson model is characterized by a traffic pattern having a variance, which approximates the mean value. As the Engset model best approximates the simulated traffic pattern (see left side of Figure 12.3), it is chosen as candidate to model traffic offered to ASONs.

As the right side of Figure 12.3 shows, the Engset model results in even a higher blocking probability, when taking into account a 95% confidence interval on the mean and variance values. This confidence interval is needed because these parameters are derived from the simulated ASON traffic pattern. The fact that the Engset model always leads to a higher blocking probability—for the same amount of available channels—would actually mean that a few channels will be overprovisioned when guaranteeing the required QoS.

Figure 12.3 suggests that designing the ASON based on the Engset model is the best choice, while not having an exact teletraffic model for ASON. Nevertheless, it is worth noting that the suitability of the Engset model in the analyzed scenario is considered under particular conditions. In fact, the model identification has been done for a small range of model parameter values (traffic among 2 and 6 Erlang and peakedness between 0.65 and 0.84). In addition, the traffic intensity involved in the system under study is of the order of few Erlangs, while in telephone networks, where such model has been successfully applied, it is on the order of tens or hundreds of Erlangs. In any case, it is most likely that just a few channels will be requested automatically in an ASON environment.

In summary, the application of Engset model for modeling and dimensioning an ASTN/ASON leads to a slight overdimensioning of the network. Nevertheless, this is a good approximation that can be used as an exact model is not available.

12.2.2 ASONs and Their Potential Benefits

The goal of this section is to illustrate the achievable benefits of ASONs by means of two sample applications. In Section 12.2.2.1, the savings for some

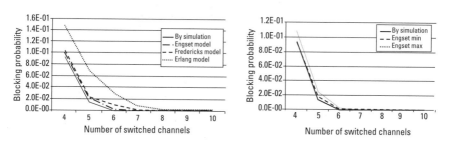

Figure 12.3 Blocking probability for an OW = b 1 min, set-up and tear-down threshold = 4kB.

emerging networking applications are illustrated, while in Section 12.2.2.2, ASONs are applied in providing multilayer survivability.

12.2.2.1 Emerging Network Applications for ASON: Opportunities and Impact on Network Resources

Within the LION project, a set of emerging applications has been investigated, including the banking sector, disaster recovery, video service delivering, grid applications, and health care services. One of the most promising candidates as potential ASON customers is the banking sector. Currently, banks are important clients of telecommunications operators. They are using permanent connections or VPNs (e.g., on-line data exchange, management, and backup). They will probably maintain at least low-capacity protected permanent connections, but the use of high-capacity switched connections would be more convenient and economical for them (especially in case of backup).

A typical bank usually consists of a headquarters, dozens of branch offices, and a few hundred local offices and automated teller machines. The whole system is guarded against severe disasters by at least one backup center. A representative bank operates on both metropolitan and core networks.

Focusing on a national core network switching STM-1, the offered load for the headquarters-branch office exchange was estimated by deriving a typical scenario for an arbitrary European country from statistics found in [4–6]. It was assumed that there are 70 banks in a hypothetical European country. All of them have headquarters in the capital. Each bank has on average 34 branch offices in the country (inclusive the four branch offices in the capital). It holds 420,000 customers' accounts of average size of about 250 KB. This means that per branch office, an amount of (420,000/34) accounts $\times$ 0.25 MB = 3.1 GB has to be transferred, which would last 2.6 min over an STM-1 circuit (= HT). Assuming that all branch offices send their data each day within 3 hours after closing, this would result in an IAT of 5.1s.

Assuming that the national backbone interconnects the capital to four main regional chief towns, the entire traffic between the capital and the rest of the country is divided into four equal parts, each one loading the main topology links. On each backbone link, we can find an offered traffic of 7.65 Erlang from the banking sector (HT = 2.6 min = 156 s, IAT = 4 $\times$ 5.1s = 20.4s) offered by 595 sources (= (34 $\times$ 70)/4). Similar assumptions for the video delivering service and health care sector results in a total traffic of 19.115 Erlang offered by 729 sources, for an average of 0.026 Erlang/source.

If we apply the Engset model to these figures, we can find that in this specific case, 32 STM-1 shared circuits are enough to assure a congestion probability lower than 0.2 %. Considering that 729 sources require in total 46 STM-16 flows, when each of them is connected through a dedicated permanent STM-1 connection, the saving ratio in this particular case is 22.8. This huge saving on one side makes the introduction of switching capabilities on transport network attractive from the point of view of network cost reduction. On the other side, it opens new perspectives in service offering. In fact, high-bandwidth connectivity services could be offered at affordable prices to some categories of customers that otherwise have to pay too much for permanent circuits in comparison to their real utilization. The significant saving in transport cost will be absorbed in part by an attractive price policy of switched services, in part by return of necessary investment in ASON-capable equipment, and in part by the constitution of new opportunities for operator revenues.

12.2.2.2 ASON for Survivability

Considering an IP/OTN multilayer network, one will traditionally provide spare capacity in advance to the IP layer in order to protect against problems (e.g., router failures). However, the ASON functionality opens opportunities to circumvent this statistical assignment of spare capacity to the IP layer by dynamically setting up or tearing down lightpaths between routers as requested by the particular situation of the moment. In this study, both static and dynamic techniques are compared in terms of physical network equipment cost [7].

In the static case, three options have been investigated. The quite simple option of double protection, where the optical layer recovery scheme also protects the IP/MPLS layer spare resources, leads to a waste of optical resources as spare capacity in the IP/MPLS layer is again protected in the optical layer. A first enhancement, avoiding double protection, is to route the spare IP/MPLS capacity as unprotected traffic over the optical layer, while the working IP/MPLS capacity is routed over the optical layer using some form of protection or restoration. Working capacity of the IP layer is thus protected in the optical layer; IP spare capacity is not. A third option (and a further improvement) is to use the common pool concept. The basic idea behind common pool is to support IP/MPLS spare resources as unprotected preemptible traffic (extra traffic) in the common pool of optical layer spare resources.

In the dynamic case, two options have been investigated: the ASON global reconfiguration strategy, in which case an optimal condition in all failure situations is calculated and assumed, and the ASON local reconfiguration strategy, which only allows capacity upgrades/downgrades on existing links. Figure 12.4 shows the results in terms of cost (relative to the nominal failure-free situation). The total network cost is split in three parts: a line cost proportional to the length of the links, a node cost proportional to the number of wavelengths entering or leaving an OXC via an aggregate port, and a tributary cost for each IP router line card connected to an OXC.

ASON local reconfiguration is clearly the most cost-efficient multilayer resilience scheme. The decreasing cost trend from *double protection* to *IP spare*

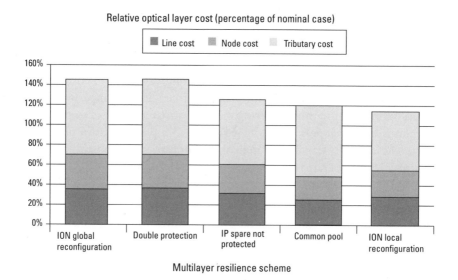

Relative optical layer cost (percentage of nominal case)

Figure 12.4 Cost comparison between static and dynamic ASON multilayer resilience schemes.

not protected to *common pool* was expected as the IP spare resources are supported more and more efficiently by the OTN resources. The higher flexibility needed to optimize the logical IP topology in each particular fault scenario in *ASON global reconfiguration* requires a higher amount of installed capacity and equipment in the optical layer than ASON local reconfiguration, making this global strategy more expensive (even as expensive as the quite inefficient static double protection strategy). The ASON local reconfiguration solution is thus less expensive than the common pool one. The main cost difference lies in the tributary cost. ASON local reconfiguration needs fewer IP router line cards. As this equipment is relatively expensive, this equipment saving results in quite a large cost saving.

12.2.3 Some Guidelines on Network Evolution

Among the emerging trends that seem to drive the evolution of current transport networks, there is the well-known increase of data-oriented applications. Some projections are showing that by the year 2003, data traffic will account for about 25% of service revenues, but it might require about 90% of the bandwidth. Nevertheless voice traffic is much more valuable than data, and the difference in value seems to increase more and more with the time. A consequence is that transport networks need to evolve in a way to drastically reduce costs and operating expenses, and to fulfill the new emerging requirements. In LION, some preliminary guidelines on the evolution of current transport networks towards next generation optical infrastructure such as ASON and GMPLS based were deducted from the obtained results.

- *Flexible provisioning of connections.* An ASON should be able to support permanent, soft-permanent, and switched connections at any time. Nevertheless, in evolutionary terms, obviously besides the permanent connections, the soft-permanent connections seem to represent the first application of network-generated signaling and routing protocols.

  If the intention is to speed up circuit provisioning, then soft-permanent circuits seem to be the first market opportunity to keep the network operator's investments on NMSs for currently deployed technology. Furthermore, from a commercial perspective, there are fewer impacts on current business models with soft-permanent circuits than with switched ones.

  As some network operators have developed their own management systems that allow circuits to be set up or taken down by means of tools within their network management systems, the move towards switched connections is likely for new technologies.

- *Types of connections.* In evolutionary terms, next generation optical network are expected to provide unidirectional and bidirectional point-to-point connections. Unidirectional point-to-multipoint connections might be provided later and can in any case be generated by means of different unidirectional requests. The capacity of the optical layer is likely to stay bidirectional in the near future. Later, however, unidirectional line systems may be introduced in the network.

- *UNI, NNI, and PNNI.* If the short-term largest market share is for soft-permanent connections, the initial emphasis should be placed on NNI development, rather than UNI development. This may also allow further examination of billing and security issues that don't need to be considered in this type of early implementation.

  Furthermore, as it seems to be feasible to run different protocols on the same network elements, network operators may be free to choose what is the most appropriate protocol suite for their needs, including the most scalable solution.

- *Restoration.* Optical restoration seems to allow cost reductions (compared with 1+1 protection, about 30% reduction of node ports), and as such it is proposed as one of the drivers for the evolution towards next generation optical networks.

  A simple and flexible multilayer recovery scheme with a limited number of reconfigurations required to work around an IP router failure and an appropriate recovery scheme in the optical layer, like ASON local reconfiguration, is an attractive solution to provide resilience in an IP/MPLS-over-OTN multilayer network.

- *Traffic models for ASON networks.* Classical teletraffic theory seems to be suitable for ASON dimensioning. It was shown that, although classical models do not lead to an exact dimensioning, they lead to a pessimistic dimensioning, which it is acceptable (better QoS than that expected).

12.3   Network Management

As previously described, the ASON concept provides two new types of transport connections, soft-permanent connections and switched connections. We call switched and soft-permanent optical connections together optical layer connections (OLCs). The new functionality is realized by enhancing the transport plane by a CP. This concept can be applied not only to OTNs, but to any type of transport network. Therefore, in the general case we call these networks ASTNs.

12.3.1   Challenges in ASON Management

The introduction of the CP into the transport network has a deep impact on the management philosophy of the new network. As there is a CP in an ASON, network resources can be configured not only by the management interface but also by the UNI. Actually, the UNI provides to the client not only the means to create and tear down OLCs; the client can also inquire the status of the connection over the UNI, and, moreover, notifications regarding the status of the connection are conveyed via the same interface.

As both management and signaling interfaces can access the same network resources, the interworking between the management and CP is an important issue that has to be properly addressed by the management concept to be developed.

The questions arising in this context can be summarized as follows:

- Concurrency between CP and management plane: How can we avoid conflicts between the CP and management plane?

- Coexistence of CP and ASTN features: What happens with the functionality implemented in the transport network, such as permanent connections? Do we still need it and, if yes, how can, for example, the scheme of permanent connection setup coexist with the soft-permanent and switched scheme?

- This leads to the more general issue of migration: How can we support, from a management point of view, the migration from classical transport networks to ASTN-enabled networks?

The next section describes the approach that LION has chosen to meet these challenges. The approach described there has been the basis for the specification of the management information model that is described in Section 12.3.3. This management interface is being implemented and tested and will be validated as an interdomain and intradomain management interface in the LION testbed. Details on the implementation of the management systems can be found in [8].

12.3.2 The LION Management Approach

The LION management approach is based on the fact that the ASTN concept will be introduced in the networks only if migration from existing transport network solutions is supported by the management concept. The reason for this is that both vendors and network operators have made large investments in the development and infrastructure of the existing transport networks and their management. Consequently, the solution should allow reusing as much as possible existing solutions and management infrastructure.

For the transport networks such as SDH, SONET, and OTN, individual management information models already exist or are under development [9–11], and the ASTN CP can be used to control each of these networks. Hence, the ASTN information model should be designed in a way that allows the reuse of the existing specifications and a combination in a modular way of the management functionality of the ASTN CP with the management functionality of either the OTN or a SONET or an SDH transport network. Therefore, the LION approach clearly separates the ASTN management functionality from the underlying transport network management functionality:

- By developing a separate information model for the CP functionality;
- By reusing for the description of the transport network the existing information models.

The separation allows the development of one general management interface for a generic ASTN that can be applied to any kind of transport network.

Furthermore, the management functionality implemented for supervision and configuration of transport networks can be reused. In particular, LION feels that the ASTN mechanisms of soft-permanent and switched setup have to coexist with permanent setup schemes. We will later show how this can be achieved by our approach.

The approach described earlier allows vendors to reuse implementations of management functionality already developed. For network operators, the advantage is that the management functionality can be aligned to well-known procedures of existing transport networks, and hence the need to adapt the operational processes can be minimized.

The next section describes the information model developed in LION for ASTN management. This description will show how the separation of ASTN management and transport network management works practically and how it influences the design of the ASTN information model.

12.3.3 Management Information Model

A management information model in general describes all information and functionality needed to manage the network. It consists of a structured set of managed objects. The managed objects represent a software abstraction of the resources to be managed and carry the management information and functionality.

For ASTNs, the main focus of management functionality is configuration and supervision of OLCs. All further management functionality is in effect important only for supporting this task. For this reason, the most important aspect of the information model is to be able to represent, from a management point of view, the OLC and its supporting resources that reside in the network element.

The management interfaces specified for existing transport networks with their different technologies [9–11] are (more or less directly) derived from a technology-independent interface specification called M.3100 [12]. This recommendation describes objects such as a *trail* (which represents the supervised end-to-end connectivity) and *connections* (which model the individual sections making up the trail). As M.3100 is a widely accepted basis for the modeling of connection-oriented networks, we have based the design of our ASTN information model on M.3100.

In Figure 12.5, the most important managed objects are depicted. This figure shows what the conceptual split into transport and CP means. As can be seen from the figure, an ASTN-enabled optical node is considered to be composed by a CP network element (CpNE) and a transport network element (TnwNE). Regarding networks, the overall network is considered to be made up of a CP network and a transport network.

The ASTN information model (highlighted in Figure 12.5 by the shaded area) only comprises resources that are part of the CP network (with other words that are directly linked to the CpNE).

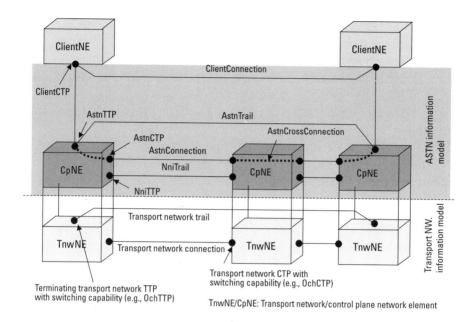

Figure 12.5 Overview on transport and CP managed resources.

From an operational perspective, the most important resource in the ASTN network is the end-to-end connectivity in the ASTN. Following M.3100, we call this resource AstnTrail.

The AstnTrail represents a switched or soft-permanent monitored OLC. In other words, it provides end-to-end data transport functionality between a pair of TnwNEs controlled by CpNEs and supervision of the quality and integrity of the signal (e.g., by mechanisms of end-to-end connectivity and continuity supervision of the CP such as LMP [13]). An AstnTrail is made up of several AstnConnections.

An AstnConnection provides data transport functionality and NNI signal transport functionality between a pair of TnwNEs controlled by CpNEs. Switched or soft-permanent OLCs (AstnTrails) are made up from several AstnConnections. Each AstnConnection object has a related transport network connection object in the respective transport network MIB.

The NniTrail provides monitored data transport functionality between a pair of TnwNEs controlled by CpNEs. In contrast to the related transport network trail object, the supervision is done on a *per link basis* using ASTN mechanisms such as those provided (e.g., by LMP [13]) rather than the trail monitoring mechanisms of the transport layer. A NniTrail acts as a server for one or more AstnConnections.

In order to be able to associate an AstnTrail to a specific client, we have introduced an object called ClientConnection, which provides end-to-end data transport functionality between a pair of ASTN client-network elements.

The trail objects mentioned here are terminated by trail termination points (TTPs); the connections are terminated by CTPs.

The AstnCrossConnection models the flexibility of the optical node and can connect AstnCTPs with each other and with AstnTTPs.

Obviously, the CP and the transport plane are not completely independent from each other. The trails configured via the CP (AstnTrails) will bind resources in the transport network and should therefore be visible to the management system of the transport network. This means that every time an AstnTrail is set up in the CP (via soft-permanent or switched scheme), the corresponding TNW trails and connections have to be created on the transport network level.

More generally it can be said that each AstnTrail, AstnConnection, and NniTrail object has a related transport network trail object in the respective transport network MIB. This is modeled by a mutual pointer relationship between these objects. By this means, which transport network resource supports which CP resource can be easily deduced. This is important for finding the root cause and to assess the impact of a failure in the transport network.

Figure 12.6 gives an overview on the structure of the LION information model as a UML diagram [14]. This representation shows which managed object is contained in which other managed object. The ControlPlaneNetwork contains the ControlPlaneNetworkElements, AstnTrails, AstnConnections, NniTrails, and ClientConnections described here.

The controlPlaneNetworkElement contains a UserNetworkInterface and networkNodeInterface to model the UNI and the NNI. These objects contain the AstnTTP and AstnTTP, where the intermediate objects describe multiplexing and link bundling. The detailed structure is described in [7, 8]. Furthermore, the ControlPlaneNetworkElement has a routingTable object which gives the management system access to the routing table. The AstnFabric models the switching fabric where the AstnCrossConnections are created. Figure 12.6 depicts further managed objects, which are related to control channel management and are not described here.

For setting up a soft-permanent OLC, the management system calls a method of the ControlPlaneNetworkElement called setupSoftpermanentOLC. In the method call, parameters such as the source and sink of the required connection UniPort objects have to be specified. In the case of multiplex capable ports (e.g., WDM ports), the channel (e.g., the wavelength) to be used inside the port also has to be specified. This information will then go into the creation of the AstnTTP objects.

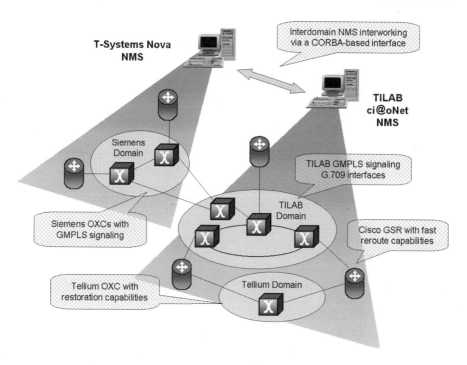

Figure 12.7 The testbed of the IST project LION.

12.4.1.1 TILAB Equipment

The equipment provided by TILAB offers transport functions at the physical level such as optical channel insertion, extraction, and switching using WDM technology. Besides, OAM&P signals can be handled for the exchange of information pertaining to:

- Network protection procedures;
- Failure identification and tracking;
- Link configuration.

These features allow for the demonstration of the main capabilities of a hierarchical, partitionable, and manageable OTN.

Both OADMs and OXCs are installed in the LION testbed. OADMs and OXCs are equipped with different line interfaces with up to eight WDM channels and different tributary ports (towards client equipment) digitally transparent up to 2.5 Gbps. Line interfaces are built on the cards that insert and extract the optical supervisory channel (OSC), while tributary interfaces are built on the transponder cards, which adapt the client signal to the transport layer. Two different kinds of tributary ports are available: 2R transponders, digitally transparent up to 2.5 Gbps, and 3R transponders with G.709 framing (digital wrapper) for client signals up to 2.5 Gbps.

All of the equipment has been realized in a previous project and has been enhanced adding the G.709 interfaces and ASON/GMPLS CP and NNI signaling. In particular, the CP being developed for the OXC is in line with the ITU-T Rec. G.8080 from which it differs for the absence of minor functionalities not related to the project objectives (i.e., mechanisms of authentication and encryption and the traffic-policing component). A more detailed description of the CP architecture is matter of a following paragraph.

12.4.1.2   SICN Equipment

TransXpress OSN managed optical distribution frame (MODIF)/OXC is the Siemens OXC platform based on microelectromechanic systems (MEMS) technology. The OXC is embedded in the MODIF. The combined TransXpress OSN MODIF/OXC concept grants an easier, faster, and manageable configuration, commissioning, and provisioning of OXCs. The core component of the TransXpress OSN MODIF/OXC solution is the OXC. It is embedded physically together with the MODIF in one system. The OXC is (like the MODIF) a transparent system, which can be used in any type of network architecture. It is bit-rate and protocol transparent and is adequate for all kind of optical signals in the 1.3 and 1.5-μm wavelength range, such as SDH/SONET, ATM, IP, and Ethernet signals, at any bit rate. The OXC switches the optical signals with the aid of optical mirrors based on MEMS. The MEMS-based switches are high-speed, low-loss optical switches (micromechanic mirrors) on silicon chips.

The switching of the OXC is controlled by an ASON controller. The ASON controller provides a SNMP interface to the TMN. A second management interface (Q3) makes it possible to manage the OXC directly without ASON functionality with the help of a local craft terminal (LCT).

Siemens provides two TransXpress OSN MODIF/OXCs, which are both configured as transparent OXCs. Each OXC has a 16 × 16-switching matrix and provides two line interfaces with three single-fiber channels each. For connecting clients, two tributary interfaces are available. The tributary interfaces are equipped with separate 2R transponders. Because the whole testbed is located in a single room, additional optical amplifiers are not necessary and are not being provided. Instead of WDM equipment, multiple fiber connections are used between the OXCs for the same reason. Additionally, for each fiber connected to an OXC, a loss of signal detector is provided that alarms a loss of signal to the ASON controller of the OXC.

12.4.1.3 Tellium Equipment

The Aurora 128 is a smaller version of Tellium's Aurora optical switch and supports 320 Gbps of bidirectional traffic for an equivalent of 128 2.5-Gbps ports or 32 10-Gbps ports. It has the same five different shelf types as the Aurora optical switch and is housed in two EIA-standard 7-ft bays. The Aurora 128 bay configuration contains one TR bay and one receiver (SR) bay.

The Aurora switches support two types of transceivers over single mode fiber: 2.5 Gbps and 10 Gbps. Both transceivers meet SONET SR specifications and can be configured in SONET or SDH format, in either channelized or concatenated mode. The Aurora switches utilize wideband receivers (1,280 nm to 1,600 nm) and standard transmit lasers on all transceivers, making wavelength translation extremely flexible and simple. The STM-16 transmitter (TR) output at 1,310 nm, and the STM-64 TRs output at 1,550 nm.

The Aurora 128 has a strictly nonblocking 128 × 128 three-stage CLOS matrix at 2.5-Gbps granularity. The entire matrix is implemented using four FTS and four MSS modules. The working and protection switch matrixes on the Aurora 128 are located in the same switch matrix shelf.

The Aurora switches provide equipment-protection switching: the working switch matrix is 1+1 protected by a protection switch matrix. The ingress transceiver bridges the incoming signals to both the working and protection switch matrixes. The two signals are independently switched and arrive at the egress transceivers. The egress transceiver selects the better of the two signals.

Network Protection

Recovery at the optical layer is fundamental to the effective design and utilization of today's high-availability core network. Optical recovery is by far the fastest recovery technology; typically, recovery at the IP layer is several orders of magnitude slower than optical layer recovery. The Aurora switches handle multiplex section protection (MSP) on the client (access) side and mesh restoration schemes via emerging standards (utilizing unused and unassigned SONET/SDH overhead bytes) on the network side. Intelligence is built into the switches to perform subsecond restoration. Connection priorities are also built into the mesh restoration algorithms.

Protection switching performance depends on the protection-switching algorithm that is being used. On the access side, Aurora switches support MSP based on G.841 (or 1+1 per GR-253-CORE) for 50-ms restoration times. On the network side, Aurora switches support StarNet dedicated mesh protection (1+1 with route diversity) for 50-ms restoration time and StarNet shared mesh restoration with a restoration objective of less than 200 ms.

- *StarNet dedicated mesh protection for both equipment and network protection.* StarNet dedicated mesh protection implements path-based 1+1 protection by providing a dedicated permanent route-diverse backup path per working path. The protection path is automatically calculated and defined during provisioning of the working path. The working and protection paths are either shared risk optical group (SROG) disjoint or node and SROG disjoint.

- *StarNet shared mesh restoration for both equipment and network protection.* StarNet shared mesh restoration implements path-based restoration by providing a shared soft-reserved route-diverse backup path per working path. The backup path is automatically calculated and defined during provisioning of the working path. The working and backup paths are either SROG disjoint or node and SROG disjoint. As opposed to StarNet dedicated mesh protection, StarNet shared mesh restoration allows sharing of backup resources because backup paths are precomputed and soft reserved. More precisely, backup paths whose working paths are route disjoint can share resources. Upon failure of the working path, the inband signaling protocol, called StarNet-P, handles the restoration and establishes the cross connects along the backup path.

Virtual Node Partitioning

In order to do multiple-nodes testing, this equipment can be logically partitioned into several virtual nodes. As with the switch, a multiplicity of processors are present (on controller boards but also on each transmitter receiver board), and every transmitter receiver board can be dedicated to a specific virtual node (the results of real-life applications, where multiple separate equipment are interconnected, are very closely approximated by this virtual node concept). For example, in the case of mesh restoration, the different boards exchange signaling messages to each other via the unused bytes in SDH overhead. As these signaling messages are processed by the TR boards, behavior of mesh restoration between TR on the same switch will be very close to the situation where TR boards are installed in different switches.

12.4.1.4   Cisco Gigabit Switch Routers

The Cisco 12000 Internet Router series offers carrier class, industry-leading scalability, high performance, reliability, and guaranteed packet delivery through an innovative distributed architecture design for service providers. Modular, multigigabit crossbar switching fabric allows 40-Gbps nonblocking switching capability. The testbed integrates five GSR Cisco 12008: this model has eight slots available for line cards and for one or two (for redundancy) gigabit route processor (GRP). The line cards adopted in the testbed have POS OC48/STM16, POS OC3/STM1, GbE, and fast Ethernet interfaces. One key feature that has been recently introduced in the MPLS-TE implementation is the fast reroute (FRR) protection, described briefly in the following paragraph.

A Brief Overview of MPLS-TE Fast Reroute

MPLS traffic-engineering FRR is a very efficient and powerful technology that allows rerouting a set of traffic engineering LSPs within a 50-ms time frame when a link/node failure occurs. FRR is a local protection mechanism; in other words, TE LSPs are locally rerouted (at the node immediately upstream to the network component failure) over a single backup tunnel (using the property of label stacking).

When a TE LSP is first signaled, every point of local repair will select the appropriate backup tunnel to use in case of link/node failure. It is worth mentioning that this selection process is done prior to any failure.

When a failure occurs and is detected, the point of local repair FRRs all the fast reroutable TE LSPs affected by the failure over the backup tunnel within 50 ms. The point of local repair also informs the various head-end LSRs of the rerouted TE LSPs that their TE LSP(s) have been locally repaired (using an RSVP path-error message) so that those head-end LSRs could trigger a reoptimization to potentially use an optimal path. The reoptimization process is non-traffic disruptive (make-before-break property): a TE LSPs is reoptimized without traffic disruption, which is an important and interesting property of TE LSPs using RSVP as the signaling protocol.

The main advantages of MPLS TE FRR can be listed as follows:

- Efficiency (50-ms recovery time);
- Scalability (a single backup tunnel is used to reroute a set of fast reroutable TE LSPs);
- Scope of failures (protects against link and layer 3 node failures);
- Granularity (recovery property defined on a per TE LSP basis);
- Fine QOS control (backup tunnel attributes allows a fine control of the QoS of the rerouted TE LSPs);
- Combination of the benefit of TE LSP reoptimization for resources optimization.

It should be noted that the failure type includes link and layer 3 node failures.

Other advanced features like backup tunnels, load balancing, and packing algorithm are also available but will not be detailed here. This technology is currently under study at the IETF.

12.4.2 Control Plane Architecture

The introduction of a CP in transport networks is likely to bring some new advantages:

- Traffic engineering for dynamic allocation of resources to routes;

- Connection control in a multivendor environment/multidomain;

- Rapid and flexible service provision;

- Introduction of supplementary and flexible optical transport services;

- Automatic optical rerouting and restoration.

ITU-T Recommendation G.8080 [1] describes the set of CP components that are used to manipulate transport network resources in order to provide the functionality of setting up, maintaining, and releasing connections.

The CP architecture is described in terms of components that represent abstract entities. Generically, every component has a set of interfaces to support a collection of operations that specify a provided or used service of that component.

Figure 12.8 shows the CP architecture according to ITU-T Recommendation G.8080 highlighting mainly the functional flows among the different components, disregarding a formal representation with all of the interfaces.

These are brief description of each component:

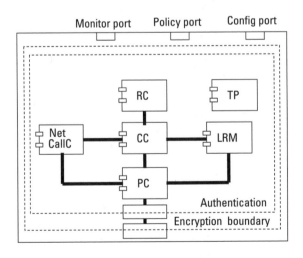

Figure 12.8 CP components.

- *CC—Connection controller component.* It manages and supervises the connection setups, releases, and the modification of connection parameters for existing connections. Moreover it is responsible for coordination among the link resource manager, routing controller, and both peer and subordinate connection controllers.

- *RC—Routing controller component.* It responds to requests from connection controllers for path information needed to set up connections and to respond to requests for topology information for network management purposes.

- *LRM—Link resource manager components.* These components are responsible for the management of an subnetwork point pool (SNPP) link, including the allocation and deallocation of link connections, provision of topology, and status information.

- *TP—Traffic policing component.* It is responsible to check that the incoming user connection is sending traffic according to the parameters.

- *NetCallC—Network call controller component.* This component accepts (after verifying user rights and resource policy) and processes incoming call requests from client networks to process and generate call termination requests towards client networks, to validate call parameters.

- *PC—Protocol controller component.* It provides the function of mapping the parameters of the abstract interfaces of the control components into messages that are carried by a protocol to support interconnection via an interface. A proper interaction between a certain number of components is necessary to control a connection.

Three approaches for dynamic path control can be identified: hierarchical, source routing, and step-by-step routing.

The hierarchical routing is based on the decomposition of a layer network into a hierarchy of subnetworks, each of them having their own dynamic connection control. A node contains a routing controller, connection controllers, and link resource managers for a single level in a subnetwork hierarchy.

In the case of source routing, the route of the connection is determined at source node. A federation of distributed connection and routing controllers implements the connection control process. The operators can specify the exact route of the path for the purpose of traffic engineering.

Step-by-step routing differs from the previous case by a reduction of routing information so that each routing controller component provides information only about the next step. In this case, the operator cannot know the route of the paths before execution of the path setup command. However, they can easily establish new paths to avoid complicated path configurations.

12.4.3　NNI Signaling for Soft-Permanent Connection Setup and Tear Down

The current NNI signaling specifications, developed within the project, utilize standards and specifications already available from other organizations or from previous internal documents and have been proposed to OIF as a contribution for the development of a standard NNI.

Within the testbed, signaling messages are exchanged over a DCN, using an out-of-fiber approach. This network should not process the signaling messages sent between the OXCs. This could be achieved in different ways (e.g., by non-RSVP-capable DCN routers, by disabling RSVP protocol in the DCN routers, by utilizing NNI messages with the router alert flag disabled, by using an RSVP implementation on the DCN routers that do not process such messages, or by using tunneling techniques between the OXCs).

Transport plane data interfaces are grouped into a *bundled link.* Both the bundled link and the component links are modeled as unnumbered interfaces according to [UNNUM] and [BUNDLE].

For soft-permanent connection, LSP setup is triggered by an NMS sending to the ingress node the appropriate information, such as source and destination node ID, port IDs, and payload type. The NMS could also specify an explicit route to the destination.

Ingress node receives this parameter directly from NMS, whereas the egress node is provided with this information by NNI signaling messages. Last subobject of the explicit route object (ERO) in the PATH message is used for this purpose. It is loaded with the node identifier and the port ID to be connected to the client node.

The teardown procedure is initiated by NMS, not independently by a node. Two different cases are possible: teardown commands sent to the source node or sent to the destination node. In both cases, NMS sends a message indicating the action to be performed (i.e., "Tear down soft-permanent LSP") and the necessary parameters (i.e., "Sender address, LSP ID, and Tunnel ID") to the involved node.

When the source node receives such a message from NMS, it sends a PathTear message downstream towards the destination node that releases reserved resources as well as the setup cross connections along the way.

If the destination node receives a "Tear down soft-permanent LSP" message from NMS, it sends a ResvTear message upstream towards the source node that releases the setup cross-connections along the way but not the reserved resources. When source node receives the ResvTear message, it will send a PathTear message to the destination node releasing the reserved resources.

At least two other possible solutions could be implemented:

1. Destination node can tear down the LSP and release the resources by a ResvTear message.

2. Destination node can request the source node to tear down the LSP and release the resources by a ResvTear message. In this case, a Resv-Tear message does not change the network state and both the cross connections and the resources will be released with the subsequent PathTear message.

12.4.4 Tests Carried Out

The experimental activities that are being carried out in the testbed are focused basically on interworking aspects investigated in the project and divided mainly into three families of tests: ASON/GMPLS functionality, network resilience, and G.709 interfaces capability.

The ASON/GMPLS experiments aim at demonstrating setup/tear down of soft-permanent connections in a multidomain and multivendor network by using NNI signaling.

The assessment of both single-layer and multilayer resilience tests is allowed by FRR capability on the IP/MPLS layer and by protection and restoration mechanisms on the ASON/GMPLS layer; the interworking between layers is guaranteed by using hold-off timers on the client layer.

Finally, the G.709 testing is focused on the analysis of the performance monitoring capabilities of these interfaces.

References

[1] ITU-T Recommendation G.8080, "Architecture for the Automatic Switched Optical Network (ASON)," Geneva, October 2001.

[2] ITU-T Recommendation G.803, "Architecture of Transport Networks Based on the Synchronous Digital Hierarchy (SDH)," 1997.

[3] ITU-T Recommendation G.872, "Architecture of Optical Transport Networks," Draft 1.4, Geneva, November 2000.

[4] National Bank of Poland, www.nbp.pl/statystyka/index.html, December 2002.

[5] Bank PBK, www.bph.pl/index_fakty.html, December 2002.

[6] Polish Offical Statistics, www.stat.gov.pl/english/index.htm, December 2002.

[7] Maesschalck, S. D., et al., "Intelligent Optical Networking for Multilayer Survivability," *IEEE Communications Magazine*, Vol. 40, No. 1, January 2002.

[8]   Lehr, G., U. Hartmer, and R. Geerdsen, "Design of a Network Level Management Information Model for Automatically Switched Transport Networks," *Network Operations and Management Symposium (NOMS)*, April 15-19, 2002, Florence, Italy.

[9]   ITU-T Recommendation G.774, "SDH Management Information Model for the Network Element View," February 2001.

[10] T1.119-1994 "Information Systems—Synchronous Optical Network (SONET)—Operations, Administration, Maintenance, and Provisioning (OAM&P) Communications," May 1994.

[11] Draft ITU-T Recommendation G.875, "Optical Transport Network (OTN) Management Information Model for the Network Element View," October 2001.

[12] ITU-T Recommendation M.3100, "Generic Network Information Model," July 1997.

[13] Lang, J. P., et al., "Link Management Protocol (LMP)," draft-ietf-mpls-lmp-02.txt, expires September 2001.

[14] Booch, G., J. Rumbaugh, and I. Jacobson, "The Unified Modeling Language User Guide," Addison Wesley, 1999.

13

Future Developments and Challenges for IP over WDM Network Deployment

13.1   Introduction

As we have pointed out in previous chapters, the future availability of IP over WDM solutions will be dictated by an evolutionary process that has already started facing the need to couple currently deployed equipment with the emerging technologies and standards. A new management model cooperating with the CP will drive a process where undoubtedly the GMPLS and the optical UNI will play key roles, as described in Chapter 12. On the other hand, in Chapters 6 through 11 we have presented a management solution for IP over WDM that is valid not only for a transitional period but also for the experience that can be gained through its assessment in different scenarios. Then, arriving at the end of the book, we want to consider the next step in this process in order to find answers to questions concerning the immediate challenges and technological developments that can pave the way.

The first section of this chapter is entirely devoted to the results of the IST-OPTIMIST initiative. This is because this project aimed at, among other objectives, the collection of technology trends, views, and roadmaps illustrating possible evolutionary scenarios in relevant areas of optical networking and highlighting possible areas for future research.

The OPTIMIST perspective is complemented with an analysis of different control/management issues. The general model of IP-over-WDM network architecture that is used as a reference for that purpose is depicted in Figure 13.1 [1]. According to this model, IP routers are connected to an optical transport network so that optical paths provide a connection between two different

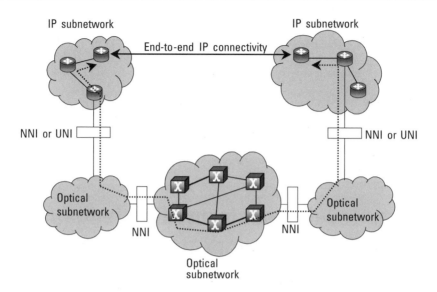

Figure 13.1 Integrated IP/WDM reference network.

routers. Physical topology of the optical layer consists of multiple optical subnetworks interconnected by optical links. Hence, the model represents a network evolved from the interconnection of different multivendor administrative domains for the provision of automatically switched end-to-end services. While interworking of the subnetworks is done by means of a standardized NNI interface, interaction between the IP routers and the optical layer is carried out either by UNI-like interface under the overlay interconnection model or by NNI under the peer and augmented models. The optical layer is responsible for the provision of point-to-point connectivity between the routers by means of fixed-bandwidth optical paths. Thus, the network virtual topology is formed by a set of lightpaths connecting source-destination pairs of IP routers. Although static at the moment, this topology is expected to become automatically reconfigurable in the near future, allowing dynamic lightpath setup as well as automatic lightpath restoration. Implementation of control protocols between the IP and the optical layers, as well as between the optical subnetworks, and their interworking with NMSs can follow different interconnection models, each representing multiple integration challenges. The most important ones are covered in the remaining sections of this chapter.

13.2   The Outlook from the IST-OPTIMIST Project

On May 1, 2000, the IST thematic network project OPTIMIST was launched. This project aims to highlight the research and development effort spent in the area of optical technologies and photonic networking in the overall European Union Fifth Framework program (FP5) IST program (1999–2002) and its contribution to the realization of the European Research Area. As the research and development effort spent on photonic technologies is scattered over several parts of the Fifth Framework program, the first aim of the OPTIMIST project was to create a thematic network actively clustering the work of the IST projects working in these technological area. The main objectives of the OPTIMIST thematic network can be summarized as follows:

- Set up a concertation process between the projects of the thematic network.
- Enhance interproject collaboration, foster synergies, and encourage consensus building.
- Support the creation and endorsement of European policies through the collection of technology trends in photonic technologies and optical networking.
- Improve the convergence between different technology platforms.
- Disseminate the results and strategies of the IST projects involved in the thematic network.
- Interact with non-IST projects/organizations at large, both at a European Union and at a worldwide level.

The OPTIMIST consortium consists of six research institutes /universities and of two consultant companies. The main activities exploited are:

- The organization of workshops/meetings/interactions within the photonic community;
- The elaboration of dissemination activities promoting the R&D results and the strategies exploited by the thematic network projects;
- The sustaining of a Web server www.ist-optimist.org acting as an entry point for everyone who is seeking more information on IST projects active in the field of optical technologies and photonic networking;
- The collection of technology trends views and the extraction of roadmaps illustrating possible evolutionary trends in the relevant area and highlighting possible areas for future research.

These types of *clustering* activities have proven to generate an added value to the European Union research programs as a whole. Concertation activities can also improve the awareness between industry and academia of new ideas and the results of longer term research in the area of photonic technologies.

Readers who wish to keep up to date with the results of IST projects active in the area of optical technologies and networking are encouraged to visit the OPTIMIST Web site or subscribe to the quarterly newsletter. These electronic services will be available for the entire OPTIMIST thematic network duration (ending May 2004), and will support the dissemination at large of the results of all of the IST projects involved.

13.2.1 Technology Trends in Optical Technologies and Photonic Networking

The identification of technology trends and possible evolutionary scenarios has been one of the major tasks of the OPTIMIST consortium members. These views and ideas are being collected during interactive workshops with the thematic network projects involved, during interaction with other photonic projects and or organizations at large, and through the study of public available research results in general. The main application domain that has been studied during the first years of the OPTIMIST project is the area of telecommunications. At the time of this writing, the application domain under study is being enlarged to other areas of industry, such as lighting and medical applications.

In the following paragraphs, however, we will only highlight the technology trends identified in the area of photonic networking and optical technologies in view of the development of the future communications infrastructure. The reader should note that although in some instances reference is made to certain IST projects, it is by no means possible within the context of this section to make a complete listing of the work carried out by all of the projects involved in the thematic network. The reader is referred to the OPTIMIST Web site for a more complete overview.

In the future, we can envisage that communication technologies will be available to all people in an invisible way, imbedded in natural surroundings, present whenever we need it, making interaction simple and effortless. This ambient intelligence (AmI) will link people through services anywhere and anytime, supporting all of our communication needs in a seamless way. Realization of this AmI environment will require an underlying network with a vast capacity. Moreover this vast connection capacity will need to be ubiquitous, flexible, reliable, and cost effective. It is clear that photonics will therefore be an essential element in the physical layer of systems and services all across the different parts of the future AmI networks. Network designers and operators indicate that it will be essential to keep the information in the fiber as long as possible. Although during the past decade, vast progress has been made in the area of photonics, many challenges still lie ahead in the realization of the fully intelligent optical network (ION). Figure 13.2 illustrates the scenario that a universal

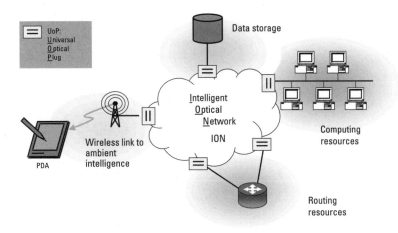

Figure 13.2 The ION providing ubiquitous connectivity.

optical plug (UoP) will be available (comparable to the electrical power grid), providing ubiquitous connectivity to the ION. Combined with appropriate, possibly nonoptical, last-mile technologies (such as wireless links as shown in the drawing) and appropriate supporting software technologies, the AmI can be realized.

As may be clear from the scenario depicted here, stringent requirements will be put on the ION. Compared to the optical transport networks of today, both the granularity of the transport capacity and the flexibility will need to be largely improved. When we depict the evolution of the OTN in the light of these characteristics, we can visualize the following scenario, as illustrated in Figure 13.3.

- At a first implementation stage, OTNs will be managed in a similar manner to the widely deployed SDH networks. This implies that these networks will be rather static, as manual interventions of the network operator are still necessary. Note also that these OTNs are circuit switched, as they cross connect high-capacity fixed bandwidth lightpipes.

- A first significant advancement in optical networking will be to automate and dynamically provide the process for setting up lightpaths. Signaling processes have to be introduced in the (optical) transport domain to realize this ASTN. These OTNs are dynamically circuit switched.

- Beyond implementing dynamic signaling functionality in the optical domain, the granularity by which the information is switched in the

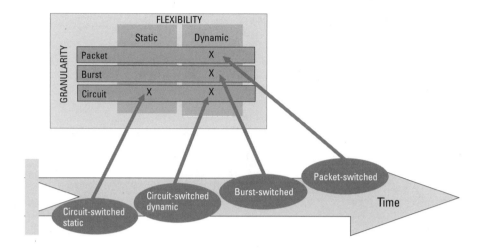

Figure 13.3 Evolution in the flexibility and in the switching granularity of the OTN.

optical domain will in the future become adaptable to the specific needs. Next to providing huge bandwidth lightpipes (circuit switching), finer switching granularities will be available as envisaged in optical burst and in optical packet switching. Switching packets optically is said to combine the advantages of statistical multiplexing and optically transiting a node (in order to avoid expensive, high-speed O/E/O conversions).

Figure 13.4 maps some of the IST projects working on photonic networking with respect to the evolution scenario depicted here.

13.2.2 The Global Communication Network in More Detail

When looking into the ION in a more detailed geographical way, we can define several parts of the network as illustrated in Figure 13.5, each with its specific characteristics, requirements, and technological trends.

The OPTIMIST consortium tried to outline some facts and figures for these several levels of the global communication network. Please note that the medium-term scenario depicts a time frame of 5 to 10 years from now; the long-term scenario looks to the possible network characteristics 10 to 20 years from now.

- The global area network (GAN), involving mainly the undersea connections, needs to provide very large connection distances without

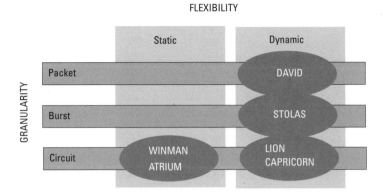

Figure 13.4 Mapping the work of some IST projects against the evolution scenarios depicted for the OTN.

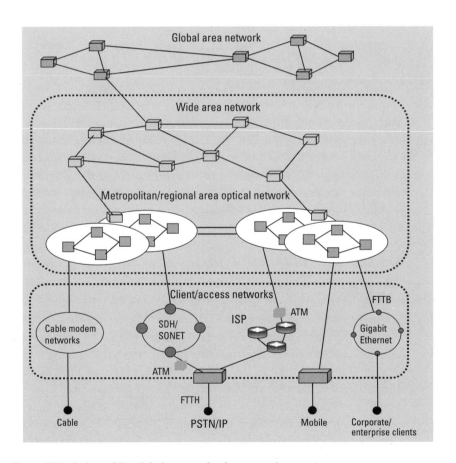

Figure 13.5 A view of the global communication network.

regeneration requirements. Because of the long distances and associated nonlinear problems, very large numbers of wavelengths are unlikely to be implemented in the GAN. In the medium-term scenario, a total transmission capacity of 4 Tbps per fiber can be depicted, carried by up to 200 channels working at 20 to 40 Gbps. On the longer term, this could evolve to 16 Tbps per fiber (up to 200 channels, 80 Gbps–100 Gbps channels speed using OTDM technology).

- A typical WAN contains spans typically ranging from hundreds to several thousands of kilometers. It interconnects networks of national size and may range over a whole continent like Europe. The nodes also serve as entry points to the metropolitan area networks (MANs) and the GAN. Within the next 10 years, WANs will evolve into meshed networks, providing optical switching at wavelength granularities. WDM with several hundreds of wavelengths will be implemented, wideband optical amplifiers, large optical switches (1,000 × 1,000 ports) and wavelength converters will be of major importance. In the long term, characteristic numbers will increase. Switching in the WAN is expected to remain on wavelength and waveband level, for this MEMS seems to be the preferred choice.

- The metropolitan area network (MAN) connects the access to the WAN/core network. It is characterized by a fairly well-defined distance scale often quoted as 20 to 200 km. Next to limited distance spans, the MAN will be characterized by high capacity needs, as it was created due to the interconnection of regional content servers. Over the long term, ultra-high speeds will be realized based on technologies such as OTDM. Due to the limited span distance, a high amount of channels seem to be feasible on the long-term scenario. The choice of technology will in the MAN area also be influenced strongly by the cost. Predicted figures are given in Table 13.1.

- The access network (AN) fans out from the access points on the metro network, connecting end users and services. Due to cost reasons, an important number of users have to share one edge node of the MAN. The AN therefore concentrates traffic streams and may also perform statistical multiplexing, thereby (in future networks) taking into account different QoSs and SLAs. The AN also has to provide the appropriate interfaces for the different services at the user sites. Due to the large impact of the legacy networks, a mix of technologies will remain to coexist in the AN. Passive optical networks (PONs) are predicted to play a major role, however, in future access networks. A predominant factor within the access domain will clearly be the cost effectiveness of the technology; very low cost, easy to handle, and

Table 13.1
Medium- and Long-Term Network Scenarios

Medium-Term Scenario (5–10 Years)	Long-Term Scenario (10–20 Years)
GAN Capacity: 4 Tbps/fiber Channel speed: 20–40 Gbps Number of wavelengths: <200 Unregenerated distance: 10,000 km OADM and switched nodes: small number	*GAN* Capacity: 16 Tbps/fiber Channel speed: 80–100 Gbps Number of wavelengths: <200 Unregenerated distance: > 10,000 km Increased number of OXCs and OADMs
WAN Capacity: 10 Tbps/fiber Channel speed: 2.5–40 Gbps Number of wavelengths: 100–256 Unregenerated distance: > 3,500 km OXC: 1,000 x 1,000	*WAN* Capacity: 20 Tbps/fiber Channel speed: 40–160 Gbps Number of wavelengths: 200–500 Unregenerated distance: > 5,000 km OXC: > 5,000 × 5,000
MAN Capacity: 10 Tbps/fiber Channel speed: 2.5 – 40 Gbps Number of wavelengths: 10–1,000 Distance: 20–200 km	*MAN* Capacity: 40 Tbps/fiber Channel speed: 10–640 Gbps Number of wavelengths: 100–256 Distance: 100–200 km

AN
Capacity:
  rings: ≤ 1 Tbps/fiber
  PONs: ≤ 1 / 250 Gbps/fiber (down/up)
Channel speed:
  meshed: ≤ 10 Gbps
  PONs: ≤ 10 / 2.5 Gbps (down/up)
Number of wavelengths: ≤ 100
Distance: ≤ 20 km
	AN
	Capacity:
	  rings: ≤10 Tbps/fiber
	  PONs: ≤ 10 / 2.5 Tbps/fiber (down/up)
	Channel speed:
	  meshed: ≤ 40 Gbps
	  PONs: ≤ 40 / 10 Gbps (down/up)
	Number of wavelengths: ≤ 250
	Distance: ≤ 40 km

robust optical components and systems will be needed in this area of the communication network.

Table 13.1 summarizes some of the data and figures collected by the OPTIMIST consortium. More detailed explanation and argumentation of these predictions and updates can be found on the OPTIMIST Web site (www.ist-optimist.org) under the section on technology trends.

13.2.3 Future Trends for Optical Components

No advanced networks without advanced components! Within the IST program, a large number of projects focus on the development of new optical materials, components, and subsystems supporting future needs and offering new possibilities to future networks. Figure 13.6 gives a schematic overview of some IST projects doing research and development work this area.

The OPTIMIST consortium analyzed with a top-down approach the needs and requirements for future communication networks and translated these requirements into development scenarios for basic components and technologies. Special focus was given to following aspects:

- Transmitter technology;

- Receiver technology;

- Fiber technology;

- Regenerator technology;

- Switching technology.

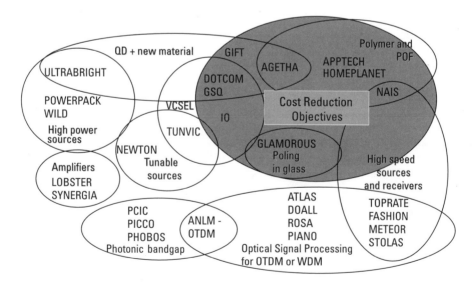

Figure 13.6 Some IST projects related to optical components and subsystems.

As an example, the diagrams given for the transmitter and for the regenerator technology are given next. More material can be found on the OPTIMIST Web site.

With respect to transmitter technology, it is obvious that WDM transmission requires stable sources locked on the International Telecommunication Union (ITU) wavelengths. Figure 13.7 maps the evolution of different types of transmitters against time for the different types of networks envisaged.

Tuneable lasers for wavelength-routed network architecture require an easy ITU-locked-wavelength allocation procedure. Tuning speed is a major issue here. Distributed Bragg reflector (DBR) lasers allow a moderate tuning range (25 nm), and the allocation procedure is fast (100 ns) but complicated and requires temperature control. External cavity lasers, on the other hand, are broadband but slow. It seems that monolithic integration of vertical cavity surface emitting lasers (VCSELs) and MEMS (movable mirrors) may lead to cost-effective solutions.

High-speed modulators are also required for high-speed signal encoding. Commercial availability of modulators and drivers at 40 Gbps and above, however, remains a critical issue.

Generation of short optical pulses is needed for OTDM interleaving and solution-managed long-haul transmission. Targets to accomplish this are transform-limited pulses or chirp-managed pulses. Solution-managed transmission or OTDM at hundreds of gigabits per second require pulse widths around

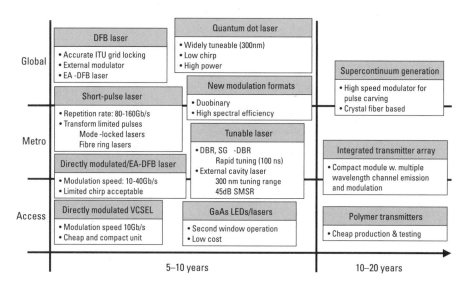

Figure 13.7 Evolution in transmitter technology.

1 ps or less. Super continuum generation in an optical fiber combined with an AWG filter can be a valid alternative to generate multiwavelength short pulses.

With respect to regenerator technology, significant research and development is still needed to reach full 3R regeneration (amplification, reshaping, retiming) in such a way that optical signals should no longer suffer from their analog behavior. Figure 13.8 depicts the evolution in regenerator technology.

For amplification, the channel width limitation is related to the amplifier coverage of the band. Several amplifier technologies can be used simultaneously: new fibers and new dopants (e.g., Tellurite fibers, Pd, and Tm doped fibers), multipump Raman amplification. Reduction of amplifier noise and cross talk are the main concerns. Nonlinear behavior in high-power fiber amplifiers is beginning to get critical. Distributed amplification is an interesting solution. Development of powerful and low-noise optical pump sources is a main issue. For metro or access links (<200 km), in-line amplifier-free connections are reliable and cost-effective solutions. Semiconductor optical amplifiers (SOAs) or planar waveguide amplifiers in silica-on-silicon may serve as low-cost boosters or preamplifiers. Signal regeneration takes place inline or at network nodes.

Successful reshaping technologies include integrated SOAs–interferometric structures, synchronized electroabsorption modulators. The technologies are highly dependent on the bit rate and thus on the choice of the multiplexing techniques (DWDM or OTDM).

For retiming of the signal clock, recovery has to be performed. Clock recovery via self-pulsating lasers and optoelectronic devices (HBT), as well as

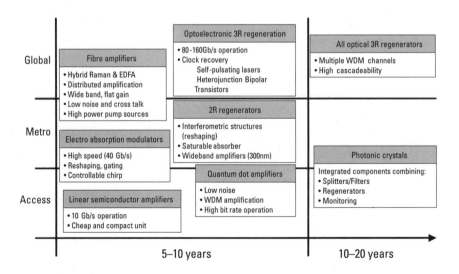

Figure 13.8 Evolution in regenerator technology.

hybrid solutions for very high OTDM rates, have been demonstrated in the laboratory.

13.3 Challenges of IP/WDM Network Control and Management

The requirement for the provision of more intelligent and autonomous networks in order to simplify and reduce the costs of network operations has caused new architectural and operational issues in all areas of network control/management. This section reviews some of the most important requirements and issues of the overlay, peer, and augmented models in the area of end-to-end network routing, wavelength routing and conversion, and fault and performance control/management. Issues associated with network security and billing/accounting are not considered here.

13.3.1 End-to-End Routing

The provision of end-to-end routing across both IP and WDM domains depends on the chosen architectural interconnection model. (See Chapter 5 for the description of network interconnection models.)

The *overlay routing* approach supports the overlay interconnection model. According to this approach, the optical network maintains a registry that allows border routers to register their IP addresses. A border router can issue a query for external addresses such that the address of an egress optical port reaching the external destination is returned.

While the deployment of the overlay model and the corresponding routing approach is likely to be the most practical one in the near future, complex management of IP routing adjacencies across the optical network is required under this model. Specifically, the WDM layer is managed by its own management system and control protocols. In order for the WDM layer to carry out the provision of optical paths on demand, it needs a separate addressing scheme and corresponding routing and signaling protocols. Additional protocols mapping an IP address to its corresponding WDM address must be developed and adopted to interconnect the IP and WDM layers. Moreover, these protocols must be responsible for discovery of IP addresses and routers across the network. Such protocols as WDM address resolution protocol (WDM-ARL) and internetworking schemes like IETF's next hop resolution protocol (NHRP) can be considered for these purposes [2].

Domain-specific routing is supported under the augmented interconnection model. Under this approach, routing in optical and IP domains are separated, and a standard protocol should be used to bridge the two domains. Border routing protocol (or, BGP) may be adapted for this purpose [1]. This protocol would allow routers to advertise their network's IP prefixes to the optical network and to receive external IP address prefixes from the optical network. Note that IP address prefixes within the optical network are not advertised to routers using BGP. Once a border OXC has received external IP prefixes from a router, it does not need to propagate the OXC address further, but it is responsible for keeping the association between external IP addresses and egress OXC addresses. When a certain external IP address is to be reached, the border router determines if a new optical path has to be established to the appropriate egress OXC. Special BGP mechanisms are to be determined for propagating egress OXC addresses.

Integrated routing is supported by the peer-interconnection model. According to this model, the same instance of an IP routing protocol is run in both IP and optical domains. The OSPF protocol can be employed with suitable optical extensions that take into account optical link parameters and any constraints specific to optical networks. For both the augmented and peer models, a uniform IP addressing scheme across both IP and WDM layers can be provided by an integrated MPLS-based control plane.

Because identical information is maintained by all OXCs and routers on both network topology and link states, a router can compute an end-to-end path to another router across the optical network. However, under the integrated routing approach, each router is responsible for maintaining a big amount of information specific the optical domain. Additionally, optical path restoration within the optical network may be visible to all nodes. Thus, integrated routing may involve significant overhead in network management. This approach may be more practical in the longer term for deployment in fast-switched or packet-switched WDM networks.

Note that regardless of the interconnection model, an IP network virtual topology achieved by interconnecting IP and WDM layers raises an issue of network scalability. The provision of a full-mesh IP virtual topology, allowing for one-hop IP connectivity (either via an IP link or via an optical path), is restricted by the number of router interfaces and OXC channels. If the full-mesh virtual topology cannot be supported (which can be the case for a network that has evolved into a large IP/WDM backbone), a multihop IP routing scheme will be necessary under the overlay model. Two layer IP MPLS over WDM MPλS stacking will be needed under the augmented and peer models, while generalized MPLS (GMPLS) architecture can be deployed to provide both packet-based and non-packet-based forwarding planes.

13.3.2 Dynamic Routing and Wavelength Conversion in the Optical Domain

The routing of optical paths in a dynamic way, inherent to the IP over reconfigurable WDM model, can be carried out by an extension of a link state routing protocol such as OSPF or IS-IS, as is used for IP routing [2]. However, in contrast to traffic-independent routing in IP layer, wavelength routing should be performed while taking into account changes in traffic load, so that path blocking due to unavailability of a free wavelength in the most congested link is minimized. High blocking of optical paths leads to dramatic increase in both packet loss rate and end-to-end delays, thus degrading performance of the whole network.

The issue of optimal routing in the optical domain is referred to as the RWA problem. While the RWA problem is shown to be MP complete [3], special heuristics such as first-fit, least-used, or most-used assignment can significantly improve the performance in terms of blocking probability [4]. Therefore, RWA computation schemes should be supported by the routing protocols. In order to distribute information on wavelength loading, the routing protocols must be extended to represent this parameter in the link state advertisement updates.

Wavelength conversion capability is another problem that affects RWA performance. On the one hand, it can reduce optical path blocking probability. However, this technology is immature and expensive at present. Thus, full wavelength conversion capability is unlikely to be deployed. Instead, limited wavelength conversion can be supported where a very small number of converters are optimally placed to minimize blocking [5]. Additionally, the converters can be placed at the age of a network administrative domain to reduce the information correlation and dependencies across the domain. Depending on the capability of the deployed cross connects, routing algorithms might have to explicitly manage wavelength continuity constraint. Thus, link state advertisement updates should support more detailed RWA information.

13.3.3 IP/WDM Fault Management

As each optical path is expected to transport of hundreds of gigabits in the near future, survivability of IP/WDM networks is a key factor to their successful deployment. Issues arising from the provision of reliable services fall into two main categories, fault localization and fault recovery.

Fault localization is one of the most difficult issues associated with IP/WDM networks with a reconfigurable optical core. In order to monitor the WDM layer, low-level analog signal parameters such as optical signal power and optical signal-to-noise ratio need to be constantly measured. It is difficult to map quantitatively these parameters to higher level IP layer observable properties such as packet loss rate. The observable IP layer outcome can be completely different, depending on a threshold value set by the NMS to generate a specific fault alarm. Additionally, the threshold values tuned to a certain bit rate may be too restrictive for lower rate signals in terms of the number of alarms generated

[2] (e.g., OC-192 may be too restrictive for an OC-48). The threshold values may also have to be updated across the network due to restoration or reconfiguration, making threshold administration very complex. For this reason, it has been proposed to include the WDM performance parameters and physical impairments in the RWA process [6].

Fault localization schemes are extremely complicated by the uncertainties intrinsic to correlating WDM-monitored signals and IP-layer parameters. Moreover, different signal processing and fault propagation properties of different WDM equipment, inevitably present in multivendor network infrastructure, add more complexity to fault localization management. Hence, a successful fault-management scheme requires tight integration across both IP and WDM domains. The development of such a scheme is under further investigation.

Network fault recovery is responsible for maintaining end-to-end IP services under network resource failures, such as link or node failures. As most currently deployed IP/WDM systems use SONET/SDH layer for fault management, new fault recovery models should be implemented when the SONET/SDH layer is eliminated. Architectures of both protection and restoration in the optical domain can be directly supported under the overlay interconnection model. While protection relies on the extra capacity allocated beforehand, restoration dynamically discovers an alternative route from spare resources in the network for the disrupted traffic once a failure has been detected. Thus, restoration architectures achieve better resource utilization with respect to protection. However, as restoration schemes need to compute the backup path after the failure has occurred, which could be carried out by means of signaling mechanisms, they lead to longer recovery times. The pros and cons of both architectures are currently under extensive investigation.

Both the augmented and the peer models with a unified GMPLS-based control plane can follow MPLS-based protection and restoration schemes, where it can be possible to set up backup lambda-switched paths. This will allow for fast switching after faults occur, so that uninterrupted connectivity is maintained. Additionally, these backup paths will be visible to the IP layer, thus providing for more efficient coordination across the two layers.

As the IP layer itself employs distributed routing algorithms designed to adapt to topology changes, it can be considered to provide IP-based fault recovery on its own. However, this approach has a few serious issues [7]. First, the slow response time of IP link failure detection might be unacceptable for the provision of real-time services. This can only be improved by modification of the IP protocol. Second, protection/restoration purely in the IP layer (i.e., fine-grain IP flow rerouting) is much less scalable compared to the WDM layer fault recovery architectures, which is the case especially for very high-capacity next generation networks. Finally, WDM layer path diversity managed by an

IP/MPLS protection/restoration algorithm is not guaranteed by using the purely IP-based approach. While an IP-based recovery might be suitable for best-effort services, its convergence with WDM protection/restoration to improve the overall efficiency of fault recovery requires further analysis.

13.3.4 IP/WDM Performance Management

Such parameters as packet loss rate, packet end-to-end delays, and packet jitter need to be bounded for QoS-capable traffic transport. The provision of premium and mission-critical services (e.g., VPNs or voice/video over IP) imposes further requirements on the ability of IP/WDM performance management to maintain SLA-guaranteed services. In particular, traffic engineering is envisaged as a means for performance optimization. IP traffic engineering is supported by MPLS, which carries out constraint-based routing in order to allow for explicit control over the IP flow paths. This way, an increased packet loss rate due to traffic pattern changes or link/node failures can be dealt with by rerouting certain LSPs away from the affected part of the network. However, it should be taken into account that the LSPs may be rerouted through nonshortest paths, resulting in increased packet end-to-end delays. Moreover, such a change in IP virtual topology may lead to highly inefficient resource utilization. Thus, the functionality of the WDM layer should be exploited to respond better to the changes in QoS performance. In particular, reconfiguration in the optical domain can be used to achieve a higher level of bandwidth utilization while maintaining the provision of QoS-capable traffic transport. WDM layer functionality should also be extended for managing differentiated services. For instance, a set of optical paths can be reserved for high-priority traffic class, or a mechanism that preempts low-priority traffic flows when high-priority traffic has to be routed over the same optical path, needs to be implemented. Additionally, physical impairments inherent to optical channels need to be taken into account when carrying out the performance management with QoS differentiation.

13.4 Deployment of IP/WDM Networks

This section considers operational features of the integrated intelligent optical CP envisaged for adoption by the service providers as well as presenting typical scenarios of near-future deployment of IP/WDM networks based on such a CP.

13.4.1 Adoption of Optical CP

At the moment, telecom operators are facing the need to couple their currently deployed equipment with the emerging networking technologies so that they can rapidly and flexibly provide existing services on demand as well as develop infrastructure for new value-added services inherent to the reconfigurable optical layer. This need inevitably leads to the development of a new network management model based on a CP that will form the bridge between the current management plane and the optical transport network itself [8]. Under this approach, GMPLS and optical UNI are considered the leaders for practical adoption. They both can be mapped into the ASON model, and they use the same underlying signaling protocols [9]. Furthermore, they have a close relationship with classical MPLS as well as common routing protocols. Hence, they are likely to become the standards for implementation of the optical CP. (See Section 5.3 for the description of GMPLS/ASON approaches.)

The development of ASON, GMPLS, and optical-UNI standards is currently progressing at a steady pace, and some vendors have already implemented support of these technologies in their products. The advantages of ASON/GMPLS deployment encompass multiple areas of IP/WDM network management. Most of the important features of IP/WDM networking functionality extended by this deployment are compared with features of the existing functionality in Table 13.2.

Automatic on-the-fly provisioning of intercarrier and multivendor-domain services, flexible service selection, and dynamic resource allocation and restoration make the approach especially attractive. However, some challenges still exist on the way to further widespread deployment of this model. They belong to such areas as further development/finalization of the standards, persuading carriers to adopt IP technology in the CP, as well as deriving the same economic benefits carriers can get by implementing optical signaling technology.

Table 13.2
Features of ASON/GMPLS–Based Networks Versus the Current IP/WDM

Existing IP/WDM Networking	ASON/GMPLS Deployment
Vendor-specific service provisioning	Intervendor service provisioning
Semimanual nonreal-time domain-by-domain provisioning	Real-time end-to-end provisioning on demand
Poor information on resource and topology	Full network inventory is maintained
Billing by ports	Billing by LSPs
Protection only	Link/path protection and dynamic restoration
Manual configuration of domains	Domains are advertised automatically

13.4.2 Possible Deployment Scenarios

From a service provider perspective, a shift towards automatically switched optical network services and the development of functionality supporting these services are dependant on the possibility of integrating the optical CP with legacy management systems. Among other issues, this raises a question of allocating the functions between the CP and management plane (e.g., whether the management plane remains responsible for carrying out routing and link management or whether these functions as well as connection processing are done by the CP).

The integration of automatically switched and currently deployed systems will be done both within different transport network layers and within transport network partitions (different vendor/carrier domains), as shown in Figure 13.9 and Figure 13.10 [10]. Consider a multivendor network whose administrative area consists of multiple control domains. Optical control plane integration can follow two steps. The first one represents a management-based solution with carrier-specific CPs directly coupled with the network management plane through necessary adaptations (Figure 13.9). While such an approach can be feasible for small-sized networks, its scalability is likely to be difficult, as multiple complex interfaces will be required for the integration when the network has expanded. Additionally, highly dynamic market business requirements make this model expensive to maintain.

As the second step, a thin layer above multiple vendor control domains will be provided as mediation between the management plane and vendor-specific domains. This layer will represent a carrier-independent common CP

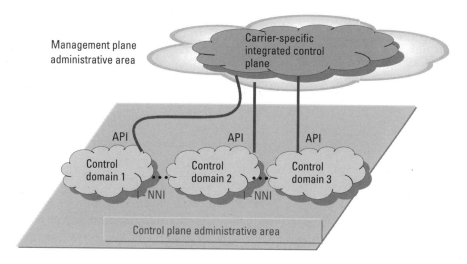

Figure 13.9 Carrier-specific CP integration scenario.

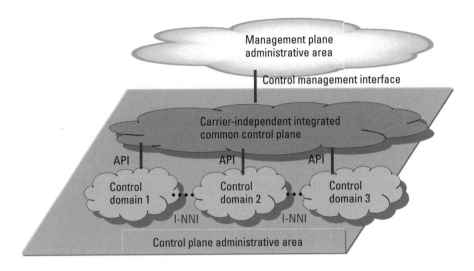

Figure 13.10 Carrier-independent CP integration scenario.

communicating with the management plane through a control-management interface (Figure 13.10). This will simplify the introduction of new administrative domains, as the integration process will be simpler and less expensive to maintain. Note that the internal NNI standard is the most probable candidate for carrying out interworking of different control domains within the same administrative area.

Consider now the integration of the optical CP (e.g., ASON) with legacy systems within different transport network partitions. New vendor domains supporting the optical CP will have to coexist with currently deployed domains that directly communicate with the management system through a transport-management interface (Figure 13.11). At the same time, the newly deployed automatically switched domains will be hidden.

This is done from the management system, by the CP interworking with the management plane through the control-management interface.

Finally, consider an implementation of interworking between different administrative domains that have both the optical CPs and legacy nonautomatically switched multiple vendor domains. These domains still directly need to be explicitly adopted to communicate with the management planes of their administrative domains. At the same time, the expected implementation of external NNI support by the optical CP will allow for smooth interoperability of both multivendor and intercarrier domains (Figure 13.12). Under this scenario, the CP and management plane will have to collaborate for the provision of intercarrier end-to-end services. While routing and link management will be carried out

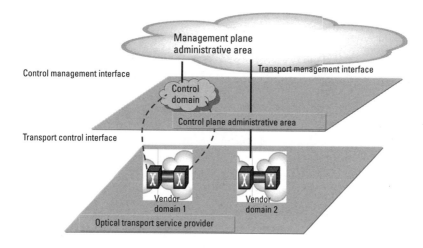

Figure 13.11 Integration of optical CP with legacy network partitions.

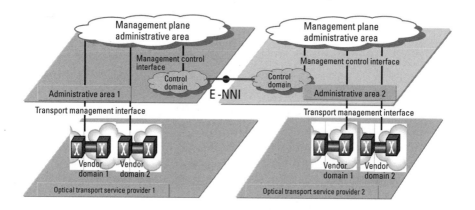

Figure 13.12 Intercarrier CPs integration.

by the management planes within each administrative domain, the E-NNI call and connection processing will be done by the optical control plane.

13.5 Summary

The IST-OPTIMIST project is aimed to highlight the research and development efforts carried out in the IST European program in the areas of optical technologies and photonic networking. More insight has been given to the prospect of the future global communication networks. The OPTIMIST project has also translated the requirement of these different types of optical networks into the requirements of the specific network devices or subsystems. Specifically, transmitters, receivers, fibers, regenerators, and switches have been covered.

Among the challenges of the control and management of IP over WDM networks, we mention dynamic routing and wavelength conversion in the optical domain. The dynamic routing of optical paths can be carried out by extensions of a link state routing protocol such as OSPF or IS-IS. However, in contrast to traffic-independent routing as in the IP layer, wavelength routing must be performed while taking into account the network traffic load. On the other hand, wavelength conversion capability is also necessary because it can reduce the optical path blocking probability. Nevertheless, this technology is as yet immature and expensive.

Fault localization is one of the most difficult issues in IP over a WDM reconfigurable optical core. This is because the mapping between the observed parameters at the optical layer, as for instance the signal-to-noise ratio, to IP-level observable properties, as the packet loss rate, is not straightforward. Additionally, the management of thresholds is very complex because threshold values tuned to specific bit rates might be not valid for other rates and because the threshold values may also have to be updated across the network due to restoration or reconfiguration processes. Fault localization schemes are also very complicated due to the intrinsic uncertainty of the correlation between monitored parameters at the IP and at the WDM layer. With respect to fault recovery, new models need to be implemented to overcome the skipping of the SDH layer that has traditionally supported this functionality.

From the service provider perspective, a shift towards automatically switched optical network services are dependent on the possibility of integrating the optical CP with legacy management systems. Optical CP integration can follow two steps. The first one represents a management-based solution with carrier-specific CPs directly coupled with the network management plane through the necessary adaptations. However, this approach does not scale well. At the second step, a thin layer above multiple vendor control domains would be provided as mediation between the management plane and vendor-specific domains. This layer would represent the carrier common CP communicating with the management plane through a control-management interface.

References

[1]   Rajagopalan, B., et al., "IP over Optical Networks: Architectural Aspects," *IEEE Commun. Mag,* September 2000, pp. 94–102.

[2]   Wei, J. Y., "Advances in the Management and Control of Optical Internet," *IEEE J. on Select. Areas Commun.,* Vol. 20, No. 4, 2002, pp. 768–785.

[3]   Baroni, S., and P. Bayvel, "Wavelength Requirements in Arbitrarily Connected Wavelength-Routed Optical Networks," *IEEE J. of Lightwave Technology,* Vol. 15, No. 2, pp. 242–251.

[4]   Zang, H., P. Jue, and B. Mukherjee, "A Review of Routing and Wavelength Assignment Approaches for Wavelength-Routed Optical WDM Networks," *Optical Networks Mag.,* January 2000, pp. 47–60.

[5]   Subramaniam, S., M. Azizoglu, and A. K. Somani, "On Optimal Converter Placement in Wavelength-Routed Networks," *IEEE/ACM Trans. Networking,* Vol. 7, No. 5, 1999, pp. 754–766.

[6]   Chiu, A., et al., "Impairments and Other Constraints on Optical Layer Routing," IETF Internet Draft,. www.ietf.org/proceedings/01aug/slides/ipo-4/tsld001.htm, November 2001.

[7]   Hjalmtysson, G., et al., "Smart Routers-Simple Optics: An Architecture for the Optical Internet," *IEEE J. Lightwave Technol.,* Vol. 18, No. 12, 2000, pp. 1880–1891.

[8]   Shahane, D., "Building Optical Control Planes: Challenges and Solutions," www.comms-design.com/story/OEG20020107S0053, January 2002.

[9]   Vissers, M., "Automatic Switched Optical Network (ASON) and Generalised MPLS (GMPLS)," www.ietf.org/proceedings/01dec/slides/plenary-2/sld001.htm, December 2001.

[10] Alanqar, W., and T. Ferris, "Carrier Motivations and Requirements for Automatically Switched Optical Network (ASON)," *ITU-T Workshop IP/Optical,* Chitose, Japan, July 9–11, 2002, www.itu.int/itudoc/itu-t/workshop/optical/s2pmp01_pp7.ppt.

List of Acronyms

AAL   ATM adaptation layer

AmI   Ambient intelligence

AN   Access network

API   Application programming interface

APS   Automatic protection switching

ASON   Automatically switched optical network

ASTN   Automatically switched transport network

ATC   ATM transfer capability

ATM   Asynchronous transfer mode

BER   Bit error rate

BGP   Border gateway protocol

BWDM   Broad wavelength division multiplexing

CAR   Committed access rate

CaSMIM   Connection and service management information model

CBR   Constant bit rate

CCAMP   Common control and measurement plane

CCD   Call clearing delay

CIM   Common information model

CLI   Command line interface

CMIP   Common management information protocol

CMISE   Common management information service element

COPS   Common open policy service

CORBA   Common object request broker

CoS   Class of service

CP   Control plane

CpNE   Control plane network element

CR-LDP   Constraint-based LDP

CS   Connectivity service

CSL   Component specification language

CTP   Connection termination point

DBR   Distributed Bragg reflector

DCN   Data communication network

DEN   Directory enabled networks

DES   Data encryption standard

DiffServ   Differentiated services

DISMAN   Distributed management working group

DOM   Document object model

DSC   Distributed software component

DTD   Document type definition

DWDM   Dense wavelength division multiplexing

E-NNI   Exterior NNI

EM   Element manager

EML   Element management layer

EMS   Element management system

ERO   Explicit route object

ES   Error seconds

FCAPS   Fault, configuration, accounting, performance, and security

FEC   Forward equivalence class

FIB   Forwarding information base

FRR   Fast reroute

FSC   Fiber-switch capable

FTN   Forward equivalence class to next hop label forwarding entry

GAN   Global area network

GbE   Gigabit Ethernet

GBIC   Gigabit interface converter

GMPLS   Generalized MPLS

GRP   Gigabit route processor

GUI   Graphical user interface

HT   Holding time

HTML   Hypertext markup language

I2   Internet2

IAT   Interarrival time

ICS   Integrated connectivity service

ID   Identifier

IDL   Interface definition language

IETF   Internet Engineering Task Force

IIOP   Internet inter-ORB protocol

ILM   Incoming label map

INMS   Interdomain network management system

I-NNI   Internal NNI

ION   Intelligent optical network

IP   Internet protocol

IPO   IP over optical

IPNM   IP network management

IPSec   IP Secure protocol

ISDN   Integrated services digital network

IS-IS   Intermediate system-intermediate system routing protocol

ISO   International Organization for Standardization

ISP   Internet service provider

IST   Information Society Technologies

ITU   International Telecommunication Union

JFC   Java foundation classes

JIDM   Joint interdomain management

LAN   Local area network

LCT   Local craft terminal

LDAP   Lightweighted directory access protocol

LDP   Label distribution protocol

LION   Layers interworking in optical networks

LOC   Loss of a single wavelength division multiplexing channel

LOH   Loss of optical overhead

LOP   Loss of payload

LOS   Loss of optical signal

LPB   Lightpath blocking

LPM   Lightpath misconnection

LSP   Label-switched path

LSR   Label-switched router

MAN   Metropolitan area network

ME   Managed element

MEMS   Microelectromechanical system

MIB   Management information base

MODIF   Managed optical distribution frame

MoIP   Multimedia over IP

MPLS   Multiprotocol label switching

MPLS-TE   Multiprotocol label switching–traffic engineering

MRTG   Multirouter traffic grapher

MSC   Message sequence chart

MSP   Multiplex section protection

MTNM   Multitechnology network management

MVC   Model-view-controller

NDM   Network data management

NGI   Next generation Internet

NHLFE   Next hop label forwarding entry

NHRP   Next hop resolution protocol

NIM   Network inventory management

NIM   Northbound interface manager

NM&R   Network maintenance and restoration

NMI-A   Network management interface–ASON

NMI-T   Network management interface–transport network

NML   Network management layer

NMS   Network management system

NNI   Network to network interface

NP   Network provisioning

OADM   Optical add-drop multiplexer

OAM   Operation and maintenance

OCh   Optical channel

OChDPRING   Optical channel dedicated protection ring

OChSPRING   Optical channel shared protection ring

OIF   Optical Interworking Forum

OLC   Optical layer connection

OLS   Optical line system

OMG   Object management group

OMS   Optical multiplexing section

OMSDPRING   Optical multiplexing section dedicated protection ring

OMSSPRING   Optical multiplexing section sharing protection ring

ONE   Optical network element

OOD   Object-oriented design

OOP   Object-oriented programming

OPTIMIST   Optical Technologies in Motion for the IST

ORB   Object request broker

OSC   Optical supervisory channel

OSMINE   Operations systems modification of intelligent network elements

O-SNCP   Optical subnetwork connection protection

OSPF   Open shortest path first

OSS   Operation support system

OTDR   Optical time domain reflectometer

OTM   Optical transport module

OTN   Optical transport network

OTS   Optical transmission section

OVPN   Optical virtual private networks

OXC   Optical cross connect

P   Provider

PBNM   Policy-based network management

PCIM   Policy core information model

PD   Performance degradation

PDD   Postdialing delay

PDH   Pleosynchronous digital hierarchy

PDP   Policy decision point

PE   Provider edge

PEP   Policy enforcement point

PF   Performance failure

PNG   Portable network graphics

POA   Portable object adapter

PON   Passive optical network

POS   Packet over SONET/SDH

POTS   Plain old telephone system

PSC   Packet-switch capable

PSTN   Public switched telephone network

PTP   Physical termination point

PVC   Permanent virtual channels

QoS   Quality of service

RADIUS   Remote authentication dial-in user service

RBOC   Regional Bell operating company

RSVP   Reservation protocol

RSVP-TE   Reservation protocol with traffic engineering extensions

RWA   Routing and wavelength assignment

SAR   Segmentation and reassembly

SAX   Simple application programming interface for extended markup language

SDL   Simple data link

SDSD   Start dial signal delay

SGML   Standard generalized markup language

SIM   Southbound interface manager

SLA   Service level agreement

SDH   Synchronous digital hierarchy

SLE   Static lightpath establishment

SLS   System level specification

SMS   Service management system

SMX   Script MIB extensibility protocol

SNC   Subnetwork connection

SNMP   Simple network management protocol

SNPP   Subnetwork point pool

SOA   Semiconductor optical amplifier

SOAP   Simple object access protocol

SOF   Shift of an optical channel frequency

SOHO   Small office home office

SONET   Synchronous optical network

SROG   Shared risk optical group

STM   Synchronous transfer mode

SVC   Switched virtual channels

TDM   Time division multiplexing

TE   Traffic engineering

TFTP   Trivial file transfer protocol

TL1   Transaction language 1

TMF   TeleManagement Forum

TMN   Telecommunication management network

TnwNE   Transport network element

TOM   Telecommunications operations map

TP   Termination point

TTP   Transmission (or trail) termination point

UBR   Unspecified bit rate

UC   Use case

UNI   User to network interface

UoP   Universal optical plug

VASP   Value-added service provider

VBR-rt   Variable bit rate—real time

VCSEL   Vertical cavity surface emitting laser

vNBS   Very-high-speed network backbone service

VoIP   Voice over IP

VPN   Virtual private network

WAN   Wide area network

WDM   Wavelength division multiplexing

WDM-ARL   WDM address resolution protocol

WFQ   Weighted fair queuing

WINMAN   WDM and IP network management

WRR   Weighted round robin

WWW   World Wide Web

xDSL   X-type digital subscriber loop

XML   Extended markup language

About the Editors

Joan Serrat received his B.A. in telecommunication engineering in 1977, and his Ph.D. in telecommunication engineering in 1983, both from the Universitat Politècnica de Catalunya (UPC). In 1977, he joined the faculty of Telecommunication Engineering, a center of UPC, where in 1986 he became an associate professor. Since then, his teaching activity has been related to communication systems engineering and his research has been in the field of signal and protocol design for data transmission networks. Since 1984, he has participated in research and development projects founded by the Spanish Public Administration (CICYT), and projects founded by Telettra Española, the European Space Agency, and the European Union. Currently, he is involved in IST FAIN, IST WINMAN, and IST CONTEXT, three European Union projects dealing with management of active networks, IP/WDM, and context-aware services, respectively. Dr. Serrat is a member of the Institution of Electrical and Electronic Engineers and the contact point for the Telemanagement Forum at UPC.

Alex Galis is a visiting professor in the Telecommunications Systems Research Group of the Department of Electronic and Electrical Engineering, University College London, England. He has published or coauthored in excess of 100 articles and reports on telecommunications subjects, including network and service management, active and programmable networks and services, intelligent and mobile agents in networks, broadband networks and global connectivity, and 3G, and beyond networks. He conducted the collaborative research and development ACTS MISA and IST FAIN projects focused on network management and active networks research. He is involved in other European Union research projects, including IST WINMAN, IST HARP, IST MANTRIP, and IST CONTEXT, which focused on network, security, and context management research.

Index

Recent Titles in the Artech House Telecommunications Library

Vinton G. Cerf, Senior Series Editor

Wide-Area Data Network Performance Engineering,
   Robert G. Cole and Ravi Ramaswamy

Winning Telco Customers Using Marketing Databases, Rob Mattison

WLANs and WPANs towards 4G Wireless, Ramjee Prasad and
   Luis Muñoz

World-Class Telecommunications Service Development, Ellen P. Ward

For further information on these and other Artech House titles,
including previously considered out-of-print books now available through our
In-Print-Forever® (IPF®) program, contact:

Artech House	Artech House
685 Canton Street	46 Gillingham Street
Norwood, MA 02062	London SW1V 1AH UK
Phone: 781-769-9750	Phone: +44 (0)20 7596-8750
Fax: 781-769-6334	Fax: +44 (0)20 7630-0166
e-mail: artech@artechhouse.com	e-mail: artech-uk@artechhouse.com

Find us on the World Wide Web at:
www.artechhouse.com